Matthias Otto

Chemometrics

⊛ WILEY-VCH

Further Reading

J. Einax, H. Zwanziger, S. Geiss
Chemometrics in Environmental Analysis
1997. ISBN 3-527-28772-8

Journal of Chemometrics
ISSN 0886-9383 (Bimonthly)

Matthias Otto

Chemometrics

Statistics and Computer Application
in Analytical Chemistry

 **WILEY-VCH**

Weinheim · New York · Chichester · Brisbane · Singapore · Toronto

Matthias Otto
Institute of Analytical Chemistry
Freiberg University of Mining
and Technology
Leipziger Str. 29
D-09599 Freiberg
Germany

Library of Congress Card No.: applied for

British Library Cataloguing-in-Publication Data:
A catalogue record for this book is available from the British Library.

Deutsche Bibliothek – Cataloguing-in-Publication Data:
Otto, Matthias:
Chemometrics : statistics and computer application in analytical chemistry / Matthias Otto. - Weinheim ; New York ; Chichester ; Brisbane ; Singapore ; Toronto : WILEY-VCH, 1999.
   Dt. Ausg. u.d.T.: Otto, Matthias: Chemometrie
   ISBN 3-527-29628-X

© WILEY-VCH Verlag GmbH, D-69469 Weinheim (Federal Republic of Germany), 1999.

Printed on acid-free and chlorine-free paper.

Composition: Satz- und Reprotechnik GmbH, D-69502 Hemsbach
Printing and Bookbinding: Druck Partner Rübelmann GmbH, D-69502 Hemsbach

Printed in the Federal Republic of Germany.

For my wife, Irmgard

Foreword

Chemometrics can be defined as "the application of mathematical and statistical techniques to chemical data". This is a broad definition. It is an appropriate definition, because chemometrics is a broad field.

In the mid 1960s, computerized instrumental methods of chemical analysis began to generate very large amounts of data. This gave rise to the so-called "data explosion" in analytical chemistry. It turned out to be an embarrassment of riches. In the past, chemists had based their decisions on only a few, hard-won, expensive, crisp pieces of data. Now they had to base their decisions on vast amounts of easily obtained, inexpensive, diffuse data. In an attempt to sort through all of this data and extract the useful, relevant information, chemists began to use mathematical and statistical techniques from other disciplines – from the social sciences especially. Chemists found that these techniques worked. They *could* extract the information they were after. And they soon discovered that these mathematical and statistical techniques gave them more than they were looking for – chemists began to discover unexpected structure in their data sets.

In the late 1960s, when I was a graduate student at Purdue, I attended Buck Roger's (required) weekly analytical chemistry seminar and heard Crawford or Morrison present their seminal work on the classification of mass spectral data. They forced their spectral data to the surface of a geometric hypersphere and made good use of the small observed distances among similar compounds, and the large observed distances among dissimilar compounds. This seminar was an "Aha!" Experience for me. It showed *me* that there was structure in chemical data! This could be useful!

Soon thereafter, Bruce Kowalski visited Purdue and gave a seminar on applications of the linear learning machine. His seminar reinforced the idea that these little-known mathematical and statistical techniques were important. About this time, Svante Wold coined the term "chemometrics", and the field had a name. With a name, the field blossomed. Many chemists, especially analytical chemists, began applying obscure mathematical and statistical techniques ("chemometric methods") to their data sets with surprisingly informative results. Experimental design was soon included as a useful chemometric tool – designed experiments are very efficient and effective for obtaining the necessary information with a minimum expenditure of resources.

In the early days, we "chemometricians" stole and tried everything we could get our hands on: principal component analysis, factor analysis, sequential simplex optimization, canonical correlation, k-nearest-neighbors, and on and on. In many cases we found appropriate applications – a good match of a mathematical or statistical technique with a particular type of chemical data set. In other cases, the match wasn't very good – in my judgement, we were trying to fit a square peg into a round hole. And we did it with vigor! For whatever reasons, we sometimes oversold these marginal methods – hyped them too much – and left it to our customers to find out that these methods weren't so good for their particular application after all. The disappointment tarnished the reputation of chemometrics. This is a shame, because chemometric methods, properly applied, are very powerful and useful methods.

Chemometrics is now a nearly mature field. The tarnish is gone. Old techniques are now being applied properly. New mathematical and statistical techniques are being developed for specific chemical applications. The results are often astounding. Chemometrics is a field worthy of study.

It is refreshing to read Matthias Otto's *Chemometrics*. Here, in one place, are clear descriptions of most chemometric methods, illustrated with relevant examples, and presented with perspective. There is no hype – only clear, reasoned, factual presentations of the methods and how they are best used.

Otto is an experienced chemometrician, a diligent scholar, and an excellent teacher. His style is crisp and clean. It is with great enthusiasm that I recommend this valuable book.

Stanley N. Deming
Houston, Texas, August 1998
Department of Chemistry
University of Houston

Preface

The idea of writing a textbook on chemometrics originates in lecturing to undergraduate and graduate students. At the Freiberg University of Mining and Technology I have given lectures and computer exercises about chemometrics for more than ten years. Further teaching has involved lecturing as visiting professor at other institutions, such as on COBAC (computer-based analytical chemistry) at the Vienna University of Technology.

From that kind of education I learned that the students work enthusiastically on the subjects of chemometrics. Initial difficulties in the computer exercises are scarcely imaginable today. Nowadays many students arrive at the university more or less as computer freaks. Problems, however, are envisaged in respect of appropriate evaluation of *chemical measurements*, because much statistical-mathematical knowledge is required. In most countries, unfortunately, basic statistical and mathematical education of chemists is poor compared, e.g., with that of physicists.

Therefore, my textbook on chemometrics teaches the most important topics of statistical-mathematical evaluation of chemical, especially analytical, measurements. It is dedicated to the evaluation of *experimental observations*, but not to theoretical aspects of chemistry.

The book is subdivided into nine chapters. In the first chapter the *subjects* of chemometrics and their *application areas* are introduced. Chapter 2 provides the *statistical fundamentals* required to describe chemical data and to apply statistical tests. The methods of *signal processing* for filtering data and for characterizing data as *time series* are the subject of Chapter 3. In Chapter 4 the methods for effective experimentation are taught on the basis of *experimental design* and *optimization*. The methods are outlined in such a way that they can be applied equally well to optimize a chemical synthesis, an analytical procedure or a drug formulation. The methods of *pattern recognition* and the assignment of data sets in the sense of *classification* are presented in Chapter 5. That chapter consists of sections on unsupervised and supervised learning. After introducing the methods of data preprocessing the typical chemometric methods for analysis of multidimensional data are outlined. Chapter 6 is dedicated to *modeling* of relationships ranging from straight-line regression to methods of multiple and nonlinear regression analysis. In Chapter 7 *analytical databanks* are discussed, i.e. the computer-accessible representation of chemical structures and spectra including the use of LIMS systems. More recent developments in chemometrics are considered in Chapter 8. Apart from the fundamentals of artificial intelligence the application of *expert systems*, of *neural networks*, of the *theory of fuzzy sets* and of *genetic algorithms* are discussed. As a very topical subject for application of statistical methods in the chemical laboratory Chapter 9 contains the most important methods for internal and external *quality assurance*, for *validation, accreditation* and for *good laboratory practice*.

In the *appendix* the reader finds statistical tables, recommendations of software, and an introduction to linear algebra. The application of chemometric methods should be eased by the use of learning objectives, by provision of approximately 60 *worked examples* and by the *questions and problems* at the end of each chapter.

The textbook is not only written for chemometric courses within the chemistry curriculum, but also for *individual study* by chemists, pharmacists, mineralogists, geologists, biologists, and scientists of related disciplines. In this context the book is considered useful also for colleagues from industry and it might be used if, e.g., multivariate methods are needed to run a NIR spectrometer, to apply statistical tests in quality assurance, or to investigate quantitative structure-activity relationships.

As usual, the text could not have been written without the comments and suggestions of my colleagues in chemometrics. My first steps in the application of computers in analytical chemistry were with Heinz Zwanziger many years ago. Later on I ran many courses on chemometrics together with Wolfhard Wegscheider. Both I name particularly on this occasion.

February 1997   Matthias Otto

Preface to the English edition

I am glad that, soon after the appearance of my book on chemometrics in German, an English edition has been finished. Thus, I had the chance to correct some mistakes and to add some methods such as the Kalman filter or a Bayesian discrimination method. Furthermore, I added learning objectives and questions and problems to the English version.

June 1998   Matthias Otto

Contents

Abbreviations

ACE	Alternating Conditional Expectations
ADU	Analog-to-Digital Converter
ANOVA	Analysis of Variance
CPU	Central Processing Unit
CRF	Chromatographic Response Function
DAC	Digital-to-Analog Converter
EPA	Environmental Protection Agency
FA	Factor Analysis
FFT	Fast Fourier Transformation
FHT	Fast Hadamard Transformation
FT	Fourier-Transformation
GC	Gas Chromatography
HORD	Hierarchically Ordered Ring Description
HOSE	Hierarchically Ordered Spherical Description of Environment
HT	Hadamard Transformation
I/O	Input/Output
IND	Indicator function
IR	Infra Red
ISO	International Organization for Standardization
JCAMP	Joint Committee on Atomic and Molecular Data
KNN	k-nearest neighbor method
LAN	Local Area Network
LDA	Linear Discriminant Analysis
LIMS	Laboratory-Information-and-Management-System
LISP	List Processing Language
LLM	Linear Learning Machine
MANOVA	Multidimensional ANOVA
MARS	Multivariate Adaptive Regression Splines
MS	Mass Spectrometry
MSDC	Mass Spectrometry Data Center
MSS	Mean Sum of Squares
NIPALS	Nonlinear Iterative Partial Least Squares
NIR	Near Infra Red
NIST	National Institute of Standards and Technology
NLR	Nonlinear Regression
NMR	Nuclear Magnetic Resonance
NPLS	Nonlinear Partial Least Squares
OLS	Ordinary Least Squares
PCA	Principal Component Analysis
PCR	Principal Component Regression
PLS	Partial Least Squares

PRESS	Predictive Residual Sum of Squares
PROLOG	Programming in Logic
RAM	Random Access Memory
RDA	Regularized Discriminant Analysis
RE	Real Error
ROM	Read Only Memory
RR	Recovery Rate
RSD	Relative Standard Deviation
RSM	Response-Surface Method
SEC	Standard Error of Calibration
SEP	Standard Error of Prediction
SIMCA	Soft Independent Modeling of Class Analogies
SS	Sum of Squares
SVD	Singular Value Decomposition
TTFA	Target-Transformation-Factor Analysis
UV	Ultraviolet
VIS	Visible

Symbols

α	Significance level (risk), separation factor
A	*Area*
b	Breadth (width), regression parameter
A^2	Quantile of chi-squared distribution
cov	Covariance
$\mathbf{C}$	Variance-covariance matrix
δ	Error
d	Distance measure, test statistic of Kolmogorov-Smirnov
D	Difference test statistic
D_k	Cook's test statistic
η	Learning coefficient
e	Error
$E(.)$	Expectation value
$\mathbf{E}$	Residual matrix
f	Degree of freedom, function
$f(x)$	Probability density function
F	Quantile of Fisher distribution
$F(\cdot)$	Function in frequency space
$f(t)$	Function in the time domain
$\mathbf{F}$	Matrix of scores
G	Geometric mean
$G(\cdot)$	Smoothing function in frequency space
$g(t)$	Smoothing function in the time domain
h	Communality
H	Harmonic mean
$\mathbf{H}$	Hat-matrix, Hadamard transformation matrix
H_0	Null hypothesis
H_1	Alternative hypothesis
$H(\cdot)$	Filter function in frequency space
$h(t)$	Filter function in the time domain
$\mathbf{I}$	Identity matrix
I_{50}	Interquartile range
$\mathbf{J}$	Jacobian matrix
k	Kurtosis
k'	Capacity factor
K_A	Protolysis (acid) constant
λ	Eigenvalue, Poisson parameter
$\mathbf{L}$	Loading matrix
μ	Population mean
m	Membership function
m_r	Moment of distribution
nf	Neighborhood function

N	Analytical resolution, plate number
$N(\nu)$	Noise
P	Probability
Q	Quartile, Dixon statistics
r	Correlation coefficient, radius
$\boldsymbol{R}$	Correlation matrix
R_s	Chromatographic resolution
R^2	Coefficient of determination
σ	Standard deviation
s	Estimate of standard deviation, skewness
s_r	Estimate of relative standard deviation
S	Similarity measure
τ	Time lag
t	Quantile of Student distribution
T	Test quantitiy of Grubb's test
$\boldsymbol{T}$	Matrix of principal components scores, transformation matrix
$\boldsymbol{U}$	Matrix of left eigenvectors
R	Range
$\boldsymbol{V}$	Matrix of right eigenvectors
w	Singular value, weight
x	(Independent) variable
$\boldsymbol{x}$	Vector of independent variables
$\boldsymbol{X}$	Matrix of independent variables
$\bar{x}$	Arithmetic mean
y	(Dependent) variable
y^*	Transformed (filtered) value
z	Standard normal deviate, signal position, objective function

1   What is chemometrics?

Learning objectives

- To define chemometrics
- To learn how to count with bits and how to perform arithmetic or logical operations with a computer
- To understand the principal terminology for computer systems and the meaning of robotics and automation

The development of the discipline chemometrics is strongly connected with the use of computers in chemistry. As early as in the seventies some analytical groups worked with statistical and mathematical methods that are ascribed nowadays to chemometric methods. Those early investigations were connected with the use mainframe computers.

The notation *Chemometrics* was introduced in 1972 by the Swede Svante Wold and the American Bruce R. Kowalski. The foundation of the International Chemometrics Society in 1974 led to the first description of this discipline. In the following years several conference series were organized, e.g. COMPANA (Computer applications in analytics), COBAC (Computer-based analytical chemistry) or CAC (Chemometrics in analytical chemistry). Some journals devoted special sections to papers on chemometric subjects. Later on novel chemometric journals were started, such as the *Journal of Chemometrics* (Wiley) and *Chemometrics and Intelligent Laboratory Systems* (Elsevier).

An actual definition of Chemometrics is:

Chemometrics is the chemical discipline that uses mathematical and statistical methods,
- *to design or select optimal measurement procedures and experiments, and*
- *to provide maximum chemical information by analyzing chemical data.*

The discipline chemometrics originates in chemistry. Typical applications of chemometric methods are the development of quantitative structure-activity relationships or the evaluation of data from chemical analysis. The data flood generated by modern analytical instrumentation is one reason, that analytical chemists in particular have developed applications of chemometric methods. Chemometrics in *analytical chemistry* is the discipline that uses mathematical and statistical methods to obtain relevant information on material systems.

1

With the availability of personal computers at the beginning of the eighties a new age commenced for acquisition, processing and interpretation of chemical data. In fact, today every scientist uses software in one form or another, related to mathematical methods or to the processing of knowledge. As a consequence the necessity emerges for a deeper understanding of those methods.

Education of chemists in mathematics and statistics is usually unsatisfactory. Therefore, one aim of chemometrics was, from the beginning, to make complicated mathematical methods practicable. Meanwhile commercial statistical and numerical software simplifies this process, so that all important chemometric methods can be taught in appropriate computer demonstrations.

Apart from the statistical-mathematical methods the topics of chemometrics are also related to problems of the computer-based laboratory, to methods for handling chemical or spectroscopic databases as well as to methods of artificial intelligence.

In addition, chemometricians contribute to the develoment of all those methods. As a rule, these developments are dedicated to special practical requirements, such as the automatic optimization of chromatographic separations or the prediction of the biological actitivity of a chemical compound.

1.1 The computer-based laboratory

Nowadays the computer is an indispensible tool in research and development. The computer is linked to analytical instrumentation; it serves as a tool for acquiring data, for word processing or for handling databases and quality-assurance systems. In addition, the computer is the basis of modern communication techniques, such as electronic mail or video conferences. In order to understand important principles of computer usage some fundamentals are considered here, i.e., coding and processing of digital information, the main components of a computer, programming languages, computer networking and automation processes.

Analog and digital data

The use of digital data has several advantages compared with the use of analog data. Digital data are less noise-sensitive. The only noise arises from round-off errors owing to finite representation of the digits of a number. They are less prone to, e.g., electrical interferences and they are compatible with digital computers.

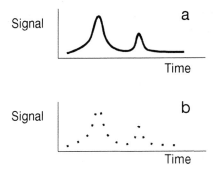

Signal | Time (a)

Signal | Time (b)

Fig. 1-1.
Dependence of signal on time for an analog (a) and a digital detector (b)

As a rule, primary data are generated as analog signals either in a discrete or a continuous mode (Fig. 1-1). For example, monitoring of the intensity of optical radiation by means of a photocell provides a continuous signal. Weak radiation, however, could be monitored by detecting individual photons by means of a photomultiplier.

Usually the analog signals generated are converted into digital data. This is performed by an analog-to-digital converter as explained below.

Binary versus decimal number system

In digital measurement the number of pulses occurring within a specified set of boundary conditions is counted. The easiest way to count is to have the pulses represented as binary numbers. In this way only two electronic states are required. To represent the decimal numbers from 0 to 9 one would need 10 different states. Typically, the binary numbers 0 and 1 are represented electronically by voltage signals of 0.5 V and 5 V, respectively. Binary numbers denote coefficients of 2 raised to different powers, so any number in the decimal system can be described.

Table 1-1.

Relationship between binary and decimal numbers

Binary number	Decimal number
0	0
1	1
10	2
11	3
100	4
101	5
110	6
111	7
1000	8
1001	9
1010	10
1101	13
10000	16
100000	32
1000000	64

Example 1-1: *Binary number representation*

The decimal number 77 is expressed as a binary number by 1001101, i.e.,

1	0	0	1	1	0	1
1×2^6	0×2^5	0×2^4	1×2^3	1×2^2	0×2^1	$1 \times 2^0 =$
64	+0	+0	+8	+4	+0	+1 = 77

Table 1-1 provides further relationships between binary and decimal numbers. Every binary number is composed of individual bits (*bit* for binary digit). The digit lying farthest to the right is termed the *least significant* digit and that furthest to the left the *most significant* digit.

How are calculations done using binary numbers? Arithmetic operations are similar, but simpler than those for decimal numbers. For addition, e.g., four combinations are feasible:

0	0	1	1
+0	+1	+0	+1
0	1	1	10

Notice that for addition of the binary numbers 1 plus 1, a 1 is carried over to the next higher power of 2.

Example 1-2: *Calculation with binary numbers*

Consider addition of 21 + 5 in decimal (a) and binary (b):

a. 21   b. 10101
    + 5   101

    26   11010

Apart from arithmetic operations in the computer, logical reasoning is also necessary. This might be in the course of an algorithm or in connection with an expert system. Logical operations with binary numbers are summarized in Table 1-2.

Table 1-2.
Truth values for logical operations on predicates p and q based on binary numbers. 1 relates to *true* and 0 represents *false*

p	q	p AND q	p OR q	IF p THEN q	NOT p
1	1	1	1	1	0
1	0	0	1	0	–
0	1	0	1	1	1
0	0	0	0	1	–

It should be mentioned that a very compact representation of numbers is based on the *hexadecimal number system*. However, hexadecimal numbers are easily converted to binary data so the details need not be explored here.

Digital and analog converters

Analog-to-digital converters (ADC)

In order to benefit from the advantages of digital data evaluation the analog signals are converted into digital ones. An analog signal consists of an infinitesimally small dense sequence of signal values in a theoretically infinite small interval. The conversion of analog into digital signals in the ADC definitely results in loss of information. For conversion, signal values are sampled at predefined time intervals and quantified in a n-ary raster (Fig. 1-2). The output signal is a code word consisting of n-bits. Using n-bits, 2^n different levels can be coded, e.g., an 8-bit-ADC has a resolution of $2^8 = 256$ amplitude levels.

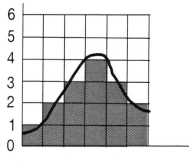

Signal

Time

Fig. 1-2.
Digitization of an analog signal by use of an analog-to-digital converter (ADC)

Digital-to-analog converters (DAC)

Converting digital into analog information is necessary if an external device must be controlled or if the data must be represented by an analog output unit. The resolution of the analog signal is determined by the number of processed bits in the converter. A 10-bit DAC provides $2^{10} = 1024$ different voltage increments. Its resolution is then 1/1024, or approximately 0.1%.

Computer terminology

Representation of numbers in a computer by *bits* has already been considered. The combination of eight bits is called a *byte*. A series of bytes arranged in sequence to represent a piece of data is termed a *word*. Typical word sizes are 8, 16, 32, or 64 bits, or 1, 2, 4, and 8 bytes.

Words are processed in *registers*. Performing a sequence of operations in a register enables *algorithms* to be performed. One or several algorithms make up a *computer program*.

The physical components of a computer form the *hardware*. Hardware includes disk and hard drives, clocks, memory units and registers for arithmetic and logical operations. Programs and instructions for the computer, including those stored on tapes and disks, represent the *software*.

Components of computers

CPU and buses

A *bus* consists of a set of parallel conductors that form the main transition lines in a computer.

The heart of a computer is the central processing unit or *CPU*. In a microprocessor or minicomputer this unit consists of a highly integrated chip.

The different components of a computer, its memory and the peripheral devices, such as printers or scanners, are joined by *buses*. To guarantee rapid communication among the various parts of a computer information is exchanged on the basis of a definitive word size, e.g., 16 bits, simultaneously over parallel lines of the bus. A data bus enables the passage of data into and out of the CPU. The origin and the destination of the data in the bus are specified by the address bus. For example an address bus with 16 lines can address $2^{16} = 65536$ different registers or other locations in the computer or in its memory. Control and status information to and from the CPU are transfered by the control bus. The peripheral devices are controlled by an external bus system, e.g., an RS 232 interface for serial data transfer or the IEEE-488 interface for parallel transfer of data.

Memory

The microcomputer or microprocessor contains typically two kinds of memory – RAM *(random access memory)* and ROM *(read only memory)*. The term RAM is somewhat misleading and historically based, since random access is feasible for RAM and ROM likewise. RAM can be used to read and write information. In constrast information in a ROM is written once, so that it can be only read but not reprogrammed. ROMs are needed in microcomputers or pocket calculators in order to perform fixed programs, e.g. for calculation of logarithms or standard deviations.

Larger programs and data collections are stored in *bulk storage devices*. At the beginning of the computer age magnetic tapes were the standard here. Nowadays tapes are still used for archiving large data amounts. Routinely 3.5 in disks (formerly 5¼ in) are used, providing a storage capacity of 1.44 Mbytes. In addition, every computer is equipped with a hard disk of at least 20 Mbyte up to several Gbyte. The access time for retrieval of stored information is of the order of a few milliseconds.

At present the availability of optical storage media is increasing. CD-ROM drives serve for reading large programs or databases. Optical hard disks can be used either to read or write information. Although optically based bulk storage devices have slower access times than magnetic bulk storage media their storage capacity is larger.

I/O systems

Communication with the computer is carried out by input-output (I/O) operations. Typical input devices are the keyboard, magnetic tapes and disks or the signals from an analytical instrument. Output devices are screens, printers, plotters, as well as tapes and disks. To convert analog information into digital or vice versa the above mentioned AD or DA converters are used.

Programs

Programming a computer with 0 and 1 states or bits is possible by use of *machine code*. Since this kind of programing is rather time-consuming higher level languages have been developed where whole groups of bit-operations are assembled. However, these so-called *assembler languages* are still difficult to handle. Therefore, high-level, algorithmic languages, such as FORTRAN, BASIC, PASCAL or C, are more common in analytical chemistry. With high-level languages the instructions for performing an algorithm can easily be formulated in a computer program. Thereafter, these instructions are translated into machine code by means of a *compiler*.

For logical programing additional high-level languages exist, e.g. LISP (list processing language) or PROLOG (programing in logic). Further developments are found in so-called *Shells*, which can be used directly for building expert systems.

Networking

Highly effective communication between computers, analytical instruments, and databases is based on networks. There are local nets, e.g., within an industrial laboratory, as well as national or worldwide networks. Local area networks (LAN) are used to transfer information about analysis samples, measure-

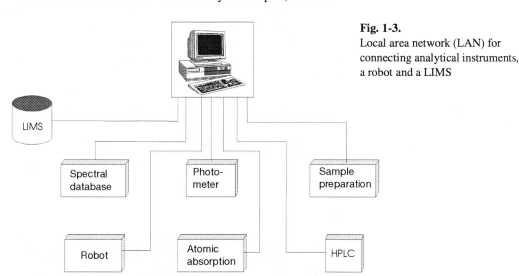

Fig. 1-3.
Local area network (LAN) for connecting analytical instruments, a robot and a LIMS

ments, research projects, or in-house databases. A typical local area network is depicted in Fig. 1-3. It contains a laboratory-and -information-management system (LIMS), by means of which all information about the sample or the progress of a project can be stored and further processed (cf. Sec. 7.1).

Worldwide networking is feasible, e.g., via the internet or CompuServe. These nets are used to exchange electronic mail (E-mail) or data with universities, research institutions, or industry.

Robotics and automation

Apart from acquiring and processing analytical data the computer can also be used to control or supervise automatic procedures. To automate manual procedures a *robot* is applied. A robot is a reprogrammable device that can perform a task more cheaply and more effectively than a person.

Typical geometric shapes of a robot arm are sketched in Fig. 1-4. The anthropomorphic geometry (Fig. 1-4A) is derived from the human torso, i.e., there is a waist, shoulder, elbow, and wrist. Although this type of robot is mainly found in the automobile industry it can also be used for manipulation of liquid or solid samples.

Fig. 1-4.
Anthropomorphic (A) and cylindrical (B) geometry of robot arms

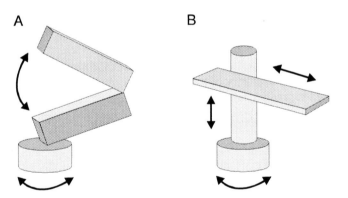

A   B

In the chemical laboratory the cylindrical geometry predominates (Fig. 1-4B). The revolving robot arm can be moved in the horizontal and vertical directions. Typical operations of a robot are:
- *Manipulation* of test tubes or glassware around the robotic work area.
- *Weighing* for determination of sample amount or for checking unit operations, e.g., addition of solvent.
- *Liquid handling* in order to dilute or add reagent solutions.
- *Conditioning* of a sample by heating or cooling.

- *Separations* based on filtration or extraction.
- *Measurements* by analytical procedures, such as spectro-photometry or chromatography.
- *Control and supervision* of different analytical steps.

Programming of a robot is based on software dedicated to the actual equipment. The software consists of elements to control the peripheral devices (robot arm, balance, pumps), to switch the devices on and off as well as to provide instructions on the basis of logical structures, e.g. IF-THEN-rules.

Alternatives for automation in a laboratory are *discrete analyzers* and *flowing systems*. By means of discrete analyzers, unit operations can be automated, such as dilution, extraction or dialysis. Continuous-flow analyzers or flow- injection analyzers serve similar objectives for automation, e.g., for the determination of clinical parameters in blood serum.

The transfer of manual operations to a robot or an automated system provides the following advantages:

1. High productivity and/or minimization of costs.
2. Improved precision and trueness of results.
3. Increased assurance for performing laboratory operations.
4. Eased validation of the different steps of an analytical procedure.

The increasing degree of automation in the laboratory leads to more and more measurements that are available on-line in the computer and have to be further processed by chemometric methods of data evaluation.

1.2 Statistics and data interpretation

Table 1-3 provides an overview of chemometric methods. The main emphasis is on statistical-mathematical methods. Random data are characterized and tested by the descriptive and inference methods of statistics, respectively. Their importance increases in connection with the aims of quality control and quality assurance. Signal processing is performed by means of algorithms for smoothing, filtering, derivation and integration. To this area belong also transformation methods such as Fourier or Hadamard transformations.

Efficient experimentation is based on the methods of experimental design and subsequent quantitative evaluation of the relationships between factors and responses. The latter can be performed by means of mathematical models or of graphical

Table 1-3.
Chemometric methods for data evaluation and interpretation

Descriptive and inference statistics
Signal processing
Experimental design
Modeling
Optimization
Pattern recognition
Classification
Artificial intelligence methods
Image processing
Information and system theory

representations. As an alternative to these simultaneous methods of experimental optimization, sequential methods are applied, such as the simplex method. There the optimum conditions are found by means of a systematic search for an objective criterion, e.g., the maximum yield of a chemical reaction, in the space of all the experimental variables.

To find patterns in data and to assign samples, materials or, in general, objects to those patterns multivariate methods of data analysis are applied. Recongition of patterns, classes or clusters is feasible with projection methods, such as principle component analysis or factor analysis, or with cluster analysis. To construct class models for classification of unknown objects we will introduce discriminant analyses.

To characterize the information content of analytical procedures information theory is used in chemometrics.

1.3 Computer-based information systems and artificial intelligence

A further subject of chemometrics is the computer-based processing of chemical structures and spectra.

It might be necessary to extract a complete or partial structure from a collection of molecular structures or to compare an unknown spectrum with the spectra in a spectral library.

For both kinds of query methods for representation and manipulation of structures and spectra in databases are needed. In addition, problems of data exchange formats, e.g. between a measured spectrum and a spectrum of a database, must be decided.

If no comparable spectrum is found in a spectral library, then methods for interpretation of spectra become necessary. For interpretation of atomic and molecular spectra all the statistical methods for pattern recognition are, in principle, appropriate (cf. Sec. 1.2). Furthermore, *methods of artificial intelligence* are also used. These include methods of logical reasoning and tools for developing expert systems. Apart from the methods of classical logic in this context, methods of approximate reasoning and of *fuzzy logic* can also be exploited. These interpretation systems constitute methods of *knowledge processing* in contrast with data processing based on mathematical-statistical methods.

Knowledge acquisition is mainly based on expert knowledge, e.g., the infrared spectroscopist is asked to contribute his knowledge in the development of an interpretation system for IR spectra. Additionally, methods are required for automatic knowledge acquisition in the form of *machine learning*.

Methods based on fuzzy theory, neural nets and evolutionary strategies are denoted *soft computing*.

The methods of artificial intelligence and machine learning are not restricted to the interpretation of spectra. They also can be used to develop expert systems, e.g., for the analysis of drugs or the synthesis of an organic compound.

Novel methods, e.g., genetic algorithms, are based on *biological analogs*, such as neural networks and evolutionary strategies. Investigation of *fractal structures* in chemistry and of models based on the *chaos* theory can be foreseen as future working areas for the chemometrician.

1.4 General reading

Sharaf, M. A., Illman, D. L., Kowalski, B. R., *Chemometrics, Chemical Analysis Series Vol. 82*: Wiley, New York, 1986.

Massart, D. L., Vandeginste, B. G. M., Deming, S. N., Buydens, L. M. C., De Jong, S., Lewi, P. J., Smeyers-Verbeke, J., *Handbook of Chemometrics and Qualimetrics, Part A and B*: Elsevier, Amsterdam, 1997 and 1998.

Questions and Problems

1. Calculate the resolution for 10-, 16- and 20-bit analog-to-digital converters.
2. How many bits are stored in a 8-bytes word?
3. What is the difference between procedural and logical programming languages?
4. Discuss typical operations of an analytical robot.

2   Basic statistics

Learning objectives

- To introduce the fundamentals of descriptive and inference statistics
- To highlight important distributions such as normal, Poisson, Student's t, F and chi-squared
- To understand the measures used to characterize the location and dispersion of a data set
- To discuss the Gaussian error propagation law
- To learn statistical tests for comparison of data sets, and for testing distributions or outliers
- To distinguish between one- and two-sided statistical tests at the lower and upper ends of a distribution
- To estimate the effect of experimental factors on the basis of univariate and multivariate analysis of variance

In analytical chemistry statistics are needed to evaluate analytical data and measurements and to preprocess, reduce and interpret the data.

As a rule analytical data are to some extent *uncertain*. There are three sources of uncertainty:
- Variability
- Measurement uncertainty
- Vagueness

A large amount of *variability* of data is typically observed with data from living beings and reflects the rich variability of nature. For example, the investigation of tissue samples provides a very variable pattern of individual compounds for each human individual.

Measurement *uncertainty* is connected with the impossibility of observing or measuring to an arbitrary level of precision and without systematic errors (bias). This is the kind of uncertainty the analyst has to consider most frequently.

Vagueness is introduced by using a natural or professional language to describe an observation, e.g., if the characterization of a property is uncertain. Typical vague descriptions are used to represent sensory variables, such as sweet taste, raspberry-colored appearance or aromatic smell.

For description of the uncertainty due to variability and measurement uncertainty statistical methods are used. Vague circumstances are characterized by fuzzy methods (cf. Sec. 8.3).

2.1 Descriptive statistics

Sources of uncertainty of analytical measurements are random and systematic errors. *Random* errors result from the limited precision of measurements. They can be reduced by replicate measurements. To characterize random errors probability based approaches are used where the measurements are considered as random, independent events.

Systematic errors (bias) represent a constant or multiplicative part of the experimental error. This error cannot be decreased by replicate measurements. In analytical chemistry the trueness of values, i.e. the deviation of the means from the true values, is related to systematic error. Appropriate measurements with standards are used to enable recognition of systematic errors in order to correct for them in later measurements.

Measurements that are dependent on each other provide correlated data. Typically, time-dependent processes, such as time series of glucose concentrations in blood, belong to that kind of data. Correlated measurements cannot be characterized by the same methods used for description of random independent observations. They require methods of time-series analysis where it is assumed that the measurements are realizations of a stochastic process and where they are statistically dependent (Sec. 3.2).

Apart from *descriptive* statistics there exists *inference* statistics (cf. Sec. 2.2).

Distribution of random numbers

In the following text we consider random and uncorrelated data. The distribution of random data can be determined from their frequency in a predefined interval also called class. As an example we consider in Table 2-1 replicate measurements of a sample solution in spectrophotometry. By partitioning the continuous variable into 12 classes one obtains the frequency of the observations in each class (Table 2-2). Graphically the frequency of observations is represented in a histogram (Fig. 2-1).

To decide on the *number of classes* n their class width w is used. The dependence of w on the number of single values n and the range R_n (Eq. (2-18)) is expressed by the relationship:

$$w = \frac{R_n}{\sqrt{n}} \quad \text{for } 30 < n \leq 400$$

$$w = \frac{R_n}{20} \quad \text{for } n > 400$$

Table 2-1.

Spectrophotometric measurements (absorbance) of a sample solution from 15 replicate measurements

Measurement	Value	Measurement	Value
1	0.3410	9	0.3430
2	0.3350	10	0.3420
3	0.3470	11	0.3560
4	0.3590	12	0.3500
5	0.3530	13	0.3630
6	0.3460	14	0.3530
7	0.3470	15	0.3480
8	0.3460		

Table 2-2.
Frequency distribution of measurements from Table 2-1

Range	Frequency	Relative Frequency, %
0.3300 to 0.3333	0	0
0.3333 to 0.3367	1	6.67
0.3367 to 0.3400	0	0
0.3400 to 0.3433	3	20.00
0.3433 to 0.3467	2	13.33
0.3467 to 0.3500	4	26.67
0.3500 to 0.3533	2	13.33
0.3533 to 0.3567	1	6.67
0.3567 to 0.3600	1	6.67
0.3600 to 0.3633	1	6.67
0.3633 to 0.3667	0	0
0.3667 to 0.3700	0	0

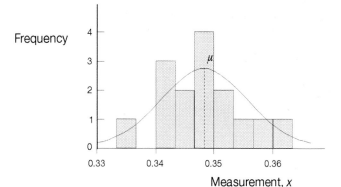

Fig. 2-1.
Histogram for the measurements of Table 2-1 and the theoretical distribution function according to a Gaussian distribution (solid line)

Gaussian distribution

If the number of replicate measurements is increased to infinity and the class width is simultaneously reduced then a bell-shaped distribution curve is obtained for the frequency of the measurements. This is called Gaussian or normal distribution and is given in Fig. 2-1 by a solid line.

The Gaussian distribution is expressed mathematicaly by:

The notions probability density function, probability function, density function and frequency function are used *synonymously*.

$$f(x) = \frac{1}{\sigma\sqrt{2\eth}}\, e^{-\frac{(x-\mu)^2}{2\sigma^2}} \qquad (2\text{-}1)$$

where

$f(x)$	–	frequency or probability function
σ	–	standard deviation
μ	–	mean
x	–	measurement (variable)

15

The mean, μ, characterizes the location of the data on the variable axis. The standard deviation, σ, and its square, the variance σ^2, describes the *dispersion* of data around the mean (cf. Fig. 2-2). The Greek letters are used by the statistician to express the true parameters of a population. Since only a limited number of measurements is available, the location and variance parameters must be estimated. The estimates are labeled by Roman letters or by use of a hat, e.g. $\hat{y}$. For estimation of the parameters of a Gaussian distribution we obtain:

Dispersion of data is also termed variation, scatter or spread.

$$f(x) = \frac{1}{s\sqrt{2\pi}} e^{-\frac{(x-\bar{x})^2}{2s^2}}$$ (2-2)

where

s — estimate of the standard deviation
$\bar{x}$ — estimate of the mean

Fig. 2-2.
Probability density distribution function of the Gaussian distribution according to Eq. 2-1 with the mean μ and the standard deviation σ

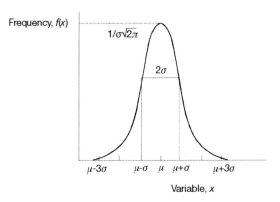

Estimation of the arithmetic mean is calculated for n replicate measurements from:

$$\bar{x} = \frac{1}{n}\sum_{i=1}^{n} x_i$$ (2-3)

The standard deviation is estimated by use of the equation:

$$s = \sqrt{\frac{\sum_{i=1}^{n}(x_i - \bar{x})^2}{n-1}}$$ (2-4)

Moments of a distribution

Mean and variance can be derived from the moments of a distribution. In general the rth central moment calculated about the mean is given by:

$$m_r(x) = \int_{-\infty}^{\infty} (x-\mu)^r f(x)\, dx$$ (2-5)

For the individual moments one obtains:

First moment: *mean*

$$m_1(x) = \int_{-\infty}^{\infty} xf(x)\,dx = \mu \qquad (2\text{-}6)$$

Second central moment: *variance*

$$m_2(x) = \int_{-\infty}^{\infty} (x-\mu)^2 f(x)\,dx = \sigma^2 \qquad (2\text{-}7)$$

Third moment: *skewness s* (asymmetry of distribution)

$$s = \frac{m_3(x)}{\sigma^3} \qquad (2\text{-}8)$$

For symmetric distributions the value of the skewness is zero. A longer tail on the right of the peak results in a skewness larger than zero. A longer tail on the left results in a skewness smaller than zero.

Fourth moment: *kurtosis k* (measure of excess)

$$k = \frac{m_4(x)}{\sigma^4} \qquad (2\text{-}9)$$

Flat and peaked distributions have values smaller and larger than zero, respectively. For the normal distribution $k = 3$, i.e. for a peaked normal distribution $k > 3$. Frequently the peakedness is defined by k-values related to the normal distribution, i.e. by the expression $k' = k - 3$.

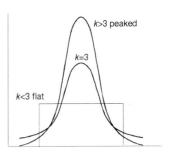

Other distributions

Apart from the distributions used for hypothesis testing, i.e. F-, t- and chi-squared distribution, to be considered in Sec. 2.2, there are further models of the distribution of random numbers. Those are the log-normal distribution, and the uniform, binomial and Poisson distributions.

The *Poisson distribution* is quite important because it enables the analyst to characterize countable, but rare events. Such events are typical for procedures based on counting rates, e.g. if a photomultiplier is applied in optical spectrometry or a proportional counter in X-ray analysis.

The Poisson distribution is based on the probability density function for discrete values of a variate. This is termed a probability function. For each value of this function, $f(x)$, a probability for the realization of the event, x, can be defined. It is

A variate specifies a random variable.

calculated according to a Poisson distribution by use of the equation:

$$f(x) = \frac{\lambda^x e^{-\lambda}}{x!} \qquad (2\text{-}10)$$

The location parameter λ determines both the mean and the variance of the distribution, i.e.:

$$\lambda = \mu = \sigma^2$$

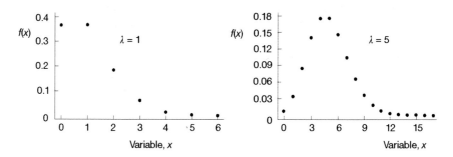

Fig. 2-3.
Probability function of the Poisson distribution according to Eq. (2-10) for two values of the parameter λ

Fig. 2-3 demonstrates the Poisson distribution for λ-values of 1 and 5, i.e. the distributions obtained when an average of 1 event or 5 events, respectively, are observed. The widths of curves, also determined by λ, are dependent only on the total number of events or the counting rate. The higher the counting rate the larger will be the variance or standard deviation of the counting process. For $x = n$ events the variance $\sigma^2 = n$ or the standard deviation for the Poisson distribution becomes:

$$\sigma = \sqrt{n} \qquad (2\text{-}11)$$

This is true for the absolute standard deviation. In contrast, the relative standard deviation is reduced as the counting rate is increased because (cf. Eq. (2-16)):

$$\sigma_r = \frac{\sigma}{n} = \frac{\sqrt{n}}{n} = \frac{1}{\sqrt{n}} \qquad (2\text{-}12)$$

Central limiting theorem

The most important distribution is the normal distribution. This conclusion can be drawn from the *central limiting theorem*:

The distribution of a sum, y, calculated from $i = 1$ for p variables, x_i (Eq. (2-13)), with means μ_i and variances σ_i^2 tends to a normal distribution with the mean $\sum_i \mu_i$ and the variance $\sum_i \sigma_i^2$, if p approaches infinity, irrespective of the distributions of the individual variables, x_i.

$$y = x_1 + x_2 + \dots x_p \qquad (2\text{-}13)$$

The central limiting theorem is illustrated in Fig. 2-4. The distribution of the considered population follows a binomial distribution. From this population several samples are drawn with 2, 4 and 25 samples in each group. Then the means of those groups are formed and plotted as the distribution of the means. Although the population investigated is binomially distributed, the distribution of the means leads to a normal distribution.

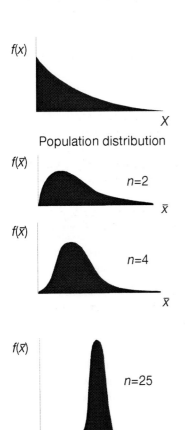

Population distribution

Sample distribution of $\bar{x}$

Fig. 2-4.
Illustration of the central limiting theorem for the distribution of means $\bar{x}$ taken from a binomial distribution X

The probability density function of a *binomial distribution* is:

$$f(x) = \binom{n}{x} p^x (1-x)^{n-x}$$

with mean np and variance $np(1 - p)$. Here n is the number of trials and p is the probability of success.

19

Location parameter

A data set can be characterized by the following quantities: *frequency, location, variance, skewness, kurtosis, quantile* and *rank*.

The only location quantity considered so far has been the arithmetic mean (Eq. (2-3)). For some problems different location parameters are more appropriate.

Geometric mean

For log-normally distributed data the geometric mean is often reported, because of the validity of the expressions:

$$G = \sqrt[n]{x_1 x_2 \dots x_n} \quad \text{or} \quad \log G = \frac{\sum_{i=1}^{n} \log x_i}{n} \tag{2-14}$$

Harmonic mean

The harmonic mean is another parameter for characterization of the most central and typical value of a set of data. Its definition is:

Median and interquartile range represent a rank order statistic. A *rank* indicates the position of an object relative to other objects by means of an ordinal number.

$$H = \frac{n}{\sum_{i=1}^{n} \left(\frac{1}{x_i} \right)} \tag{2-15}$$

When the geometric mean exists, it lies between the harmonic and arithmetic means, i.e. $H \leq G \leq \bar{x}$.

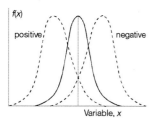

For symmetric distributions the mean, $\bar{x}$, the median and the mode are identical.

For *positive* values of the skewness: $\bar{x} >$ median $>$ mode;

For *negative* skewness: $\bar{x} <$ median $<$ mode.

Median

A very robust location measure is the median. For an odd number of values the median is the middle-order statistic. It lies at the position $(n + 1)/2$. For an even number of measurements the median is calculated from the average of the $(n/2)$th and $(n/2 + 1)$th-order statistics. The median is less dependent on outliers than is the arithmetic mean.

Quartile

Looking from a different perspective the median represents that point that divides the total frequency into two halves. i.e. 50% of the data are found above and below the median, respectively. This point is also termed the middle quartile $Q(0.5)$ or Q_2. The quartiles Q_1 and Q_3 are called the lower and upper quartiles, respectively. The lower quartile contains 25% of all measurements, the upper 75%. A percentile $p\%$ divides the data range into hundreds.

Dispersion measures

As the most common dispersion measure we have already used the standard deviation. Frequently this measure is reported as the *relative standard deviation*:

$$s_r = \frac{s}{x} \qquad \text{or, in percent,} \qquad s_r (\%) = s_r \, 100 \qquad (2\text{-}16)$$

The *mode* is the most frequently occurring value in a set of data. For categorized data the mode is equivalent to the class with the highest frequency.

This quantity is also termed the *coefficient of variation*.

The *standard error* characterizes the averaged error of the mean of n observations:

$$s_{\bar{x}} = \frac{s}{\sqrt{n}} \qquad (2\text{-}17)$$

The difference between the maximum, x_{max}, and minimum, x_{min}, values is termed the *range R*:

$$R = x_{max} - x_{min} \qquad (2\text{-}18)$$

This quantity describes the range which contains 100% of all observations. If a range that contains just 50% of all the observations is of interest, then the *interquartile range* must be calculated:

$$I_{50} = Q_3 - Q_1 \qquad (2\text{-}19)$$

The interquartile range is obtained from the difference between the lower and upper quartiles.

A *quantile* divides a set of observations into two groups, such that one fraction falls above and the complementary fraction below the value specified by the quantile. The most frequently applied quantiles are quartiles and percentiles.

Confidence interval

The confidence interval characterizes the range about the mean of a random variable in which an observation can be expected with a given probability P or risk $\alpha = 1 - P$. As statistical factor the t-value from Student's distribution is used (cf. Sec. 2.2). The confidence interval for the mean, $\bar{x}$, is calculated for f degrees of freedom from:

$$\Delta x = t(1 - \alpha/2; f) \cdot s_{\bar{x}} \qquad (2\text{-}20)$$

Inserting the standard deviation of the mean according to Eq. (2-17) gives:

$$\Delta x = \frac{t(1 - \alpha/2; f) \cdot s}{\sqrt{n}} \qquad (2\text{-}20\,a)$$

The following example provides an overview of the discussed quantities used for descriptive statistics.

Example 2-1: *Descriptive statistics*

Table 2-3 summarizes the values calculated for the descriptive statistics for the spectrophotometric data given in Table 2-1.

Table 2-3.
Descriptive statistics for the spectrophotometric measurements in Table 2-1

Sample number, n	15
Arithmetic mean, $\bar{x}$	0.3486
Median	0.347
Geometric mean, G	0.3485
Variance, s^2	0.000053
Standard deviation, s	0.00731
Relative standard deviation., $\%s_r$	2.096
Standard error, $s_{\bar{x}}$	0.00189
Confidence interval, Δx at $\alpha = 0.05$	0.00405
Minimum value, x_{min}	0.335
Maximum value, x_{max}	0.363
Range, R	0.028
Lower quartile, Q_1	0.343
Upper quartile, Q_3	0.353
Interquartile range, I_{50}	0.01
Skewness	0.259
Kurtosis	0.0452

Descriptive statistics graphically illustrated: box-whisker-plots

Graphically the most important details of the descriptive statistics of data can be represented as a box-whisker-plot or, for short, box-plot (Fig. 2-5). Along the variable axis, here the ordinate, a box is drawn, with the lower and upper quartiles being the bottom and top of the box, respectively. The width of the box has no significance.

Fig. 2-5.
Box-whisker-plot for the data in Table 2-1 with an additional outlier at an absorbance of 0.373

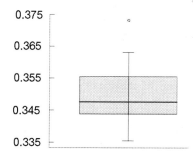

The whiskers are obtained as follows: the upper adjacent value is equal to the interquartile range plus 1.5 times that distance:

Upper adjacent value $= Q_3 + 1.5(Q_3 - Q_1)$ (2-21)

The lower adjacent value is equal to the lower quartile minus 1.5 times the interquartile range:

Lower adjacent value $= Q_1 - 1.5(Q_3 - Q_1)$ (2-22)

Values outside the adjacent values are considered outliers. They are plotted as individual points. Box-whisker-plots are not restricted to illustrating the univariate statistics of a single variable. Plots of several variables enables their different distribution characteristics to be compared easily.

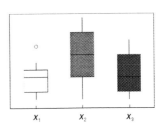

Error propagation

The uncertainty of analytical measurements originates from different sources. Among these are:
- Sampling
- Instrumental deviations, e.g. a wrongly calibrated balance
- Reagent impurities
- Measurement conditions (e.g. influence of temperature or humidity)
- Matrix effects
- Round-off errors
- Contamination from the environment or between individual samples
- Operator effects
- Random influences

In order to estimate the uncertainty of analytical results the propagation of error cannot be considered alone for a single effect, e.g. the reproducibility of instrumental measurement, but *all* sources of uncertainty must be taken into account for all steps of the analytical procedure.

The uncertainty of an analytical result is obtained from the Gaussian law of error propagation. Assuming a general analytical observation, y, that is dependent on m factors x_i according to a function f:

$$y = f(x_1, x_2, ..., x_m)$$ (2-23)

The uncertainty of the final result based on the deviation from the mean can then be described in terms of the dependence on the uncertainty of the factors by:

$$dy = df(x_1, x_2, ..., x_m)$$ (2-24)

As a local approximation of the unknown function f the partial derivatives with respect to all factors are formed by calculating the total differentials:

$$dy = \left(\frac{\ddot{a}y}{\delta x_1}\right)_{x_2, \ldots, x_m} dx_1 + \left(\frac{\delta y}{\delta x_2}\right)_{x_1, \ldots, x_m} \times dx_2 \ldots + \left(\frac{\delta y}{\delta x_m}\right)_{x_1, x_2, \ldots} dx_m$$

(2-25)

To link this quantity to the variance, the deviations are squared on both sides of Eq. (2-25):

$$(dy)^2 = \left[\left(\frac{\delta y}{\delta x_1}\right)_{x_2, \ldots, x_m} dx_1 + \left(\frac{\delta y}{\delta x_2}\right)_{x_1, \ldots, x_m} \times dx_2 \ldots + \left(\frac{\delta y}{\delta x_m}\right)_{x_1, x_2, \ldots} dx_m\right]^2$$

(2-26)

In squaring Eq. (2-26) two types of term emerge from the right hand side of the equation: square terms and cross terms, e.g.

$$\left(\frac{\delta y}{\delta x_1}\right)^2 dx_1^2 \quad \text{and} \quad \left(\frac{\delta y}{\delta x_1}\right)\left(\frac{\delta y}{\delta x_2}\right) dx_1 dx_2$$

The square terms must always be considered, since they come out as positive terms irrespective of the sign of the partial derivatives. The cross terms can be either positively or negatively signed. As long as the factors are independent of each other the cross terms will approximately cancel out and so can be neglected. If the factors are dependent on each other then the cross terms must be included in calculating the uncertainty of the whole procedure.

Computation of the uncertainty on the basis of the variances is carried out after appropriate reshaping (normalization to $n - 1$ measurements) without accounting for the crossed terms according to:

$$s_y^2 = \left(\frac{\delta y}{\delta x_1}\right)^2 s_{x_1}^2 + \left(\frac{\delta y}{\delta x_2}\right)^2 s_{x_2}^2 \ldots + \left(\frac{\delta y}{\delta x_m}\right)^2 s_{x_m}^2$$

(2-27)

Frequently the uncertainty is given by its standard deviation s_y, i.e. the square root of Eq. (2-27).

The propagation of error is exemplified in Table 2-4 for typical cases. As can be seen from Table 2-4, the variance of the observation y changes because of its dependence on the *abso-*

lute values of the variances of the individual quantities if the factors obey an additive or subtractive relationship. For multiplicative or divisive relationships the final variance is determined by the *relative* variances.

For reporting of an error interval with a given statistical certainty Eq. (2-27) must be multiplied by a factor k, e.g. $k = 2$ for a 95% probability (cf. Table 2-5) to find the measurements in this interval.

Table 2-4.
Examples of error propagation for different dependences of the analytical observation, y, on the factors, x

Relationship	Calculation of uncertainty
$\left.\begin{array}{l} y = x_1 + x_2 \\ y = x_1 - x_2 \end{array}\right\}$	$s_y^2 = s_{x_1}^2 + s_{x_2}^2$
$\left.\begin{array}{l} y = x_1 \cdot x_2 \\ y = x_1 / x_2 \end{array}\right\}$	$\dfrac{s_y^2}{y^2} = \left(\dfrac{s_{x_1}}{x_1}\right)^2 + \left(\dfrac{s_{x_2}}{x_2}\right)^2$
$y = x^a$	$\dfrac{s_y^2}{y^2} = \left(a\dfrac{s_x}{x}\right)^2$
$y = \log_{10} x$	$s_y^2 = \left(0.434\dfrac{s_x}{x}\right)^2$
$y = \text{anti}\log_{10} x$	$\dfrac{s_y^2}{y^2} = (2.303 s_x)^2$

Uncertainty and error

At the end of this section the difference between error and uncertainty will be stated. The *error* describes the difference between a measured quantity and the true or expected value. It is expressed as a single value for a given measurement. In principle, an error can be corrected for.

The *uncertainty* of a result characterizes a range and is valid for a group of measurements or for all measurements considered. Correction of uncertainty is basically not feasible.

2.2 Statistical tests

In the previous section we used statistics only for the description of data. In many cases it is necessary, however, to draw conclusions from comparisons of data at a given statistical significance. These test methods belong to the inference methods of statistics. For testing hypotheses we need to learn about some more distributions, such as the t-, F- and χ^2-distribution.

The standard normal distribution

For testing hypotheses the probability density function of the Gaussian distribution is standardized to an area of 1 and a mean of 0. This is done by introducing a standardized variable *(the standard normal variate)*, z, which expresses the deviation of the observations x from the mean μ related to the standard deviation σ:

$$z = \frac{x - \mu}{\sigma} \tag{2-28}$$

Transforming all x-values into values of the deviate z results in a density function of the *standard normal distribution* according to the model:

$$f(z) = \frac{1}{\sqrt{2\pi}} e^{-\frac{z^2}{2}} \tag{2-29}$$

As already apparent from Fig. 2-2 the probability of occurrence of an observation decreases with increasing variance or standard deviation σ. The probability that an observation is contained within a given variance range is denoted by P (cf. Fig. 2-6).

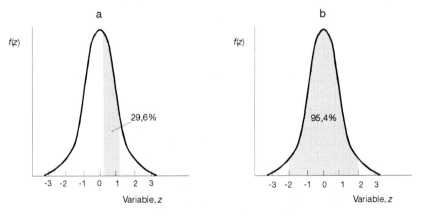

Fig. 2-6. Examples of the integration of the Gaussian distribution in the ranges for the deviate z from 0.25 to 1.25 (a) and for $z = \pm 2$ (b)

This probability can also be derived from the distribution curve – the error integral $F(x)$ – of the Gaussian distribution (Fig. 2-7). An analogous consideration leads to the risk $\alpha = 1 - P$.

Important ranges for the error integral are given in Table 2-5 in connection with the related percentage and tail areas in units of the standard deviation.

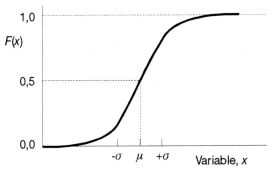

Table 2-5.
Important areas according to the error integral

Limits at the x-axis		Fraction of the total area P in %	Tail area α in %
x_1	x_2		
$\mu - 1\sigma$	$\mu + 1\sigma$	68.3	31.7
$\mu - 2\sigma$	$\mu + 2\sigma$	95.4	4.6
$\mu - 3\sigma$	$\mu + 3\sigma$	99.7	0.3
$\mu - 1.96\sigma$	$\mu + 1.96\sigma$	95	5
$\mu - 2.58\sigma$	$\mu + 2.58\sigma$	99	1
$\mu - 3.29\sigma$	$\mu + 3.29\sigma$	99.9	0.1

Example 2-2: *Normal distribution*

Determination of phenol in waste water revealed normally distributed values with a mean of $\mu = 0.6$ µg L^{-1} and a standard deviation of $\sigma = 0.04$ µg L^{-1}. How large is the probability that in a subsequent measurement the phenol concentration is contained in the range between 0.61 and 0.65 µg L^{-1}?

According to Eq. (2-28) the value of the deviate z is calculated at the upper und lower limits by:

$$z_1 = \frac{0.61 - 0.6}{0.04} = 0.25 \quad z_2 = \frac{0.65 - 0.6}{0.04} = 1.25$$

For the areas under the Gaussian curve one obtains from Table II in the appendix (the curve covers values between 0 and +z) $F(z_1) = 0.0987$ and $F(z_2) = 0.3944$. The difference between both areas provides the probability of the occurrence of the measurements, i.e. the probability is $0.3944 - 0.0987 = 0.296$ or 29.6 % (cf. Fig. 2-6a).

Especially important areas under the Gaussian curve correspond to probabilities of 95% and 99% and the complementary risk values of 5% and 1% (cf. Table 2-5). They are commonly used as significance levels in hypothesis testing.

The probability-based considerations above serve as the most important fundamentals for derivation of statistically assured decisions. In general in inference statistics the first step is the definition of a hypothesis and the significance of which is tested against a given risk α.

Testing hypotheses

In general, the test of a hypothesis consists of five steps. First, the null hypothesis H_0 and alternative hypothesis H_1 are defined. Second a test statistic must be chosen. Third the significance level must be specified. Fourth a decision rule must be set up that is based on the significance level and the distribution of the test statistics. Fifth the test statistic from the sample must be calculated and the decision on the decision rule must be made.

For example, the *null hypothesis* might postulate the randomness of samples in a group of observations. If the Null hypothesis H_0 is rejected the *alternative hypothesis* H_1 must be accepted. Since with practical measurements only a limited number of samples of the population will be available, the statistical tests usually cannot be directly based on the Gaussian distribution, but have to be performed on distributions derived from the normal distribution.

Comparison of a mean with a true value: one-variable *t*-test

Testing a sample mean $\bar{x}$ obtained in an experimental measurement against a population mean μ from a normal distribution is carried out on the basis of a Gaussian- or *t*-test. In analytical chemistry this is applied for comparison of an experimental mean with a true value.

The significance level represents the probability that the null hypothesis is falsely rejected.

The *null hypothesis* (H_0) reads: both samples belong to the same population, i.e. the difference between the sample and the true value is random, $\bar{x} = \mu$.

When the null hypothesis is rejected the *alternative hypothesis* (H_1) is valid, $\bar{x} \neq \mu$, which suggests that the sample mean is different from the true value.

Before performing the test the significance level is defined. Typically the risk α is chosen to be 0.05 or 0.01 (cf. Table 2-5).

When the standard deviation, σ, of the population is known the testing procedure can be based on the Gaussian test. The test statistic z is calculated for n parallel measurements from:

$$z = \frac{|\bar{x} - \mu|}{\sigma} \sqrt{n} \qquad (2\text{-}30)$$

Comparison of z with the quantile of the standard normal distribution (Table II in the appendix) gives the result: if z is smaller than or equal to the quantile for a given risk level α and the number of degrees of freedom is $f = n - 1$, then the null hypothesis must be accepted. Otherwise, if z is larger than z, the alternative hypothesis is accepted.

In practice the population standard deviation σ is usually unknown. Then comparison of a sample mean with the true mean must be based on Student's t-test. The test statistic is derived from Student's t-distribution as follows:

$$t = \frac{|\bar{x} - \mu|}{s} \sqrt{n}$$ (2-31)

with s – estimate of the standard deviation
n – number of parallel measurements

Comparison of the calculated test value t with the tabulated value of the t-distribution (Table 2-6) at a given risk level α for the degree of freedom $f = n - 1$, $t(1 - \alpha/2; f)$, provides the decision on the test.

Table 2-6.
Quantile of the one-sided Student's t-distribution for three significance levels α and different degrees of freedom f.
Note how the distribution aproaches the Gaussian distribution if the degrees of freedom approaches infinity (cf. Table 2-5)

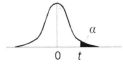

f	$\alpha = 0.05$	$\alpha = 0.025$	$\alpha = 0.01$	f	$\alpha = 0.05$	$\alpha = 0.025$	$\alpha = 0.01$
1	6.314	12.706	31.821	21	1.721	2.080	2.518
2	2.920	4.303	6.965	22	1.717	2.074	2.508
3	2.353	3.182	4.541	23	1.714	2.069	2.500
4	2.132	2.776	3.747	24	1.711	2.064	2.492
5	2.015	2.571	3.365	25	1.708	2.060	2.485
6	1.943	2.447	3.143	26	1.706	2.056	2.479
7	1.895	2.365	2.998	27	1.703	2.052	2.473
8	1.860	2.306	2.896	28	1.701	2.048	2.467
9	1.833	2.262	2.821	29	1.699	2.045	2.462
10	1.812	2.228	2.764	30	1.697	2.042	2.457
11	1.796	2.201	2.718	40	1.684	2.021	2.423
12	1.782	2.179	2.681	50	1.676	2.009	2.403
13	1.771	2.160	2.650	60	1.671	2.000	2.390
14	1.761	2.145	2.624	70	1.667	1.994	2.381
15	1.753	2.131	2.602	80	1.664	1.990	2.374
16	1.746	2.120	2.583	90	1.662	1.987	2.369
17	1.740	2.110	2.567	100	1.660	1.984	2.364
18	1.734	2.101	2.552	300	1.650	1.968	2.339
19	1.729	2.093	2.539	800	1.647	1.963	2.331
20	1.725	2.086	2.528		1.645	1.960	2.326

When $t \leq t(1 - \alpha/2; f)$ the result of the test is not significant, i.e. the null hypothesis ($\bar{x} = \mu$) must be accepted. In other words the sample mean is only randomly different from the true one. For $t > t(1 - \alpha/2; f)$ the test shows that the results are indicative of statistical significance and the alternative hypothesis is accepted ($\bar{x} \neq \mu$).

Nonparametric (distribution-free) tests do not require assumptions about the distribution of the population of the features to be tested. An example is the Wilcoxon test.

In the above example a *two-sided* t-test was applied. A *one-sided* test is valid if, for example, it has to be tested whether a sample mean $\bar{x}$ follows a regulated value μ at a significance level α. This question is tested for by the following steps:

Null hypothesis H_0: $\bar{x} \leq \mu$

Alternative hypothesis H_1: $\bar{x} > \mu$

The tabulated value of the t-distribution is taken at the upper end one-sided, i.e. the value is $t(1 - \alpha; f)$. If $t < t(1 - \alpha; f)$, then the result of the statistical test is significant and the regulated value is covered.

The tabulated t-value at an α-level of 0.05 at the upper end of the one-sided t-distribution is $t\,(1 - \alpha = 0.95; f = 3) = 2.353$. Because $t > t\,(1 - \alpha = 0.95; f = 3)$, the null hypothesis is rejected and the alternative hypothesis is accepted, i.e. the sample mean is greater than the limiting value.

The analogous question, whether a regulated value of 50 mg L^{-1} nitrate is exceeded leads also to a one-sided test. The hypotheses are:

Null hypothesis H_0: $\bar{x} \geq \mu$

Alternative hypothesis H_1: $\bar{x} < \mu$

The critical value is now to be taken at the lower end of the distribution, i.e. $t\,(\alpha = 0.05; f = 3) = -2.353$ and the null hypothesis is to be accepted, if $t > t\,(\alpha; f)$. Since by calculation according to Eq. (2-31) $t > t\,(\alpha = 0.05; f = 3)$ the null hypothesis is accepted, i.e. the regulated value is indeed exceeded.

Note: Although the used tabulated t-value of -2.353 is not given in Table 2-6 it can be derived from the table since:

$$t\,(1 - \alpha; f) = -t\,(\alpha; f) \qquad (2\text{-}33)$$

Also it must be mentioned that in the case of the one-sided test in Eq. (2-31) the actual value and not the absolute value must be used. To understand the different hypothesis tests in more detail they are summarized in Table 2-7 and illustrated in Fig. 2-8.

The principles of one- and two-sided tests are applied for tests based on other distributions.

Table 2-7.
Overview on hypothesis testing based on Student's t-test

t-test	Null hypothesis	Alternative hypothesis	t-value for the acception of H_0	Fig.
One-sided at the upper end	H_0: $\bar{x} \leq \mu$	H_1: $\bar{x} > \mu$	$t < t\,(1 - \alpha; f = n - 1)$	2-8a
One-sided at the lower end	H_0: $\bar{x} \geq \mu$	H_1: $\bar{x} < \mu$	$t > t\,(\alpha; f = n - 1)$	2-8b
Two-sided	H_0: $\bar{x} = \mu$	H_1: $\bar{x} \neq \mu$	$|t| < t\,(1 - \alpha/2; f = n - 1)$	2-8c

p-level instead of comparing test quantities

In statistical software it is rather unusual to carry out statistical tests the way we have done it above. The programs usually compute that level of significance at which the test is actually rejected. This value is called p-level or *level attained*. Is the p-level lower than or equal to the adjusted level, then the null hypothesis is rejected, otherwise it is accepted.

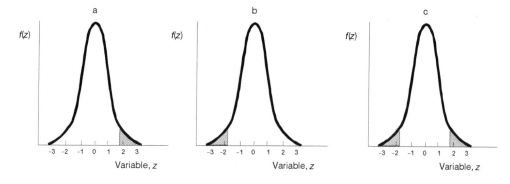

Fig. 2.8. Illustration of critical areas for one-sided tests at the upper (a) and lower (b) ends, and for two-sided test

> **Example 2-5:** *p-Level*
>
> In example 2-4 we evaluated the nitrate concentration of drinking water using a one-sided *t*-test at the upper end and tested the hypothesis that the regulated value of 50 mg L^{-1} nitrate is observed (H_0: $\bar{x} \leq \mu$, H_1: $\bar{x} > \mu$). The computed *t*-value corresponds to a significance level (*p*-level) of 0.002373. This value is lower than the specified level of $\alpha = 0.05$, so that the null hypothesis is rejected as before.

Comparison of two means: two-variable *t*-tests

A statistical test only makes sense if the significance level is fixed in advance, even if the *p*-level can be calculated after the test at any precision.

A comparison of two sample means $\bar{x}_1$ and $\bar{x}_2$ is done as follows:

$$t = \frac{|\bar{x}_1 - \bar{x}_2|}{s_d} \sqrt{\frac{n_1 n_2}{n_1 + n_2}} \qquad (2\text{-}34)$$

where

n_1, n_2 – number of parallel determinations for the means $\bar{x}_1$ and $\bar{x}_2$

s_d – weighted averaged standard deviation:

$$s_d = \sqrt{\frac{(n_1 - 1)s_1^2 + (n_2 - 1)s_2^2}{n_1 + n_2 - 2}}$$

The null hypothesis is accepted if the two means $\bar{x}_1$ and $\bar{x}_2$ are different only randomly at risk level α, i.e. if the calculated *t*-value is lower than the tabulated value for t $(1 - \alpha/2; f = n_1 + n_2 - 2)$.

The assumption for this so-called *extended t*-test is the comparability of the variances of the two random samples s_1^2 und s_2^2. Comparability means here, that the two variances are equal at a given statistical significance level. The significance of the differences between the two variances is tested by means of an *F*-test (see below).

In case the differences between the variances are not negligible, the general t-test, e.g. after Welch, has to be applied:

$$t = \frac{|\bar{x}_1 - \bar{x}_2|}{\sqrt{\dfrac{s_1^2}{n_1} + \dfrac{s_2^2}{n_2}}} \qquad (2\text{-}35)$$

The number of degrees of freedom f for the test statistics t $(P; f)$ is calculated according to the equation:

$$f = \frac{\left(\dfrac{s_1^2}{n_1} + \dfrac{s_2^2}{n_2}\right)^2}{\dfrac{\left(s_1^2 / n_1\right)^2}{n_1 - 1} + \dfrac{\left(s_2^2 / n_2\right)^2}{n_2 - 1}} \qquad (2\text{-}36)$$

Table 2-8.
F-Quantiles for $\alpha = 0.05$ (normal) and $\alpha = 0.01$ (bold) and for different degrees of freedom f_1 and f_2

f_2	$f_1 = 1$	2	3	4	5	6	7	8	9	10
1	161	200	216	225	230	234	237	239	241	242
	4052	**4999**	**5403**	**5625**	**5764**	**5859**	**5928**	**5981**	**6022**	**6056**
2	18.51	19.00	19.16	19.25	19.30	19.33	19.36	19.37	19.38	19.39
	98.49	**9900**	**99.17**	**99.25**	**99.30**	**99.33**	**99.36**	**99.37**	**99.39**	**99.40**
3	10.13	9.55	9.28	9.12	9.01	8.94	8.88	8.84	8.81	8.78
	34.12	**30.82**	**29.46**	**28.71**	**28.24**	**27.91**	**27.67**	**27.49**	**27.34**	**27.23**
4	7.71	6.94	6.59	6.39	6.26	6.16	6.09	6.04	6.00	5.96
	21.20	**18.00**	**16.69**	**15.98**	**15.52**	**15.21**	**14.98**	**14.80**	**14.66**	**14.54**
5	6.61	5.79	5.41	5.19	5.05	4.95	4.88	4.82	4.78	4.74
	16.26	**13.27**	**12.06**	**11.39**	**10.97**	**10.67**	**10.45**	**10.29**	**10.15**	**10.05**
6	5.99	5.14	4.76	4.53	4.39	4.28	4.21	4.15	4.10	4.06
	13.74	**10.92**	**9.78**	**9.15**	**8.75**	**8.47**	**8.26**	**8.10**	**7.98**	**7.87**
7	5.59	4.74	4.35	4.12	3.97	3.87	3.79	3.73	3.68	3.63
	12.25	**9.55**	**8.45**	**7.85**	**7.46**	**7.19**	**7.00**	**6.84**	**6.71**	**6.62**
8	5.32	4.46	4.07	3.84	3.69	3.58	3.50	3.44	3.39	3.34
	11.26	**8.65**	**7.59**	**7.01**	**6.63**	**6.37**	**6.19**	**6.03**	**5.91**	**5.82**
9	5.12	4.26	3.86	3.63	3.48	3.37	3.29	3.23	3.18	3.13
	10.56	**8.02**	**6.99**	**6.42**	**6.06**	**5.80**	**5.62**	**5.47**	**5.35**	**5.26**
10	4.96	4.10	3.71	3.48	3.33	3.22	3.14	3.07	3.02	2.97
	10.04	**7.56**	**6.55**	**5.99**	**5.64**	**5.39**	**5.21**	**5.06**	**4.95**	**4.85**

33

Comparison of variances: _F_-test

To compare the variances of two random samples or their standard deviations Fischer's _F_-test is applied. The _F_-value is calculated from the variances s_1^2 and s_2^2 by use of equation:

$$F = \frac{s_1^2}{s_2^2} \quad \text{(where } s_1^2 > s_2^2 \text{)} \tag{2-37}$$

The null hypothesis is accepted if the variances s_1^2 and s_2^2 differ only randomly, i.e. if the calculated _F_-value is lower than the value of the _F_-distribution at risk level α and the number of degrees of freedom $f_1 = n_1 - 1$ and $f_2 = n_2 - 1$. Tabulated values of the _F_-distribution at significance levels of 0.95 and 0.99 are given in Table 2-8.

Example 2-6: _Extended t-test and F-test_

The titanium content of steel is determined in two laboratories by means of atomic absorption spectrometry. The data are given in the table in the margin. After estimation of the variances of the determinations in the two labs the means should be compared on the basis of the appropriate _t_-test.

To compare the variances the _F_-test is carried out acccording to Eq. (2-37). On the basis of the standard deviations for:

Lab 1: $s_1 = 0.0229$ and

Lab 2: $s_2 = 0.0182$

the following _F_-value from the corresponding variances is obtained:

$$F = \frac{s_1^2}{s_2^2} = \frac{0.0229^2}{0.0182^2} = 1.58$$

The critical _F_-value is taken from Table IV to be:

$$F(1 - \alpha/2; f_1, f_2) = F(0.975; 7, 5) = 6.85$$

The calculated _F_-value is lower than that tabulated, i.e. the test result is not significant and the variances differ only randomly.

To compare the means the extended two-sided _t_-test for _comparable variances_ can be applied according to Eq. (2-34). Using the mean values:

Lab 1: $\bar{x}_1 = 0.467$

Lab 2: $\bar{x}_2 = 0.503$

Determination of _titanium contents_ (absolute %) by two laboratories

Lab 1	Lab 2
0.470	0.529
0.448	0.490
0.463	0.489
0.449	0.521
0.482	0.486
0.454	0.502
0.477	
0.409	

and the above mentioned standard deviations we calculate that:

$$s_d = \sqrt{\frac{(8-1)0.0299^2 + (6-1)0.0182^2}{8+6-2}} = 0.0211$$

for the test quantity:

$$t = \frac{|\bar{x}_1 - \bar{x}_2|}{s_d} \sqrt{\frac{n_1 n_2}{n_1 + n_2}} = \frac{|0.467 - 0.503|}{0.0211} \sqrt{\frac{6 \cdot 8}{6 + 8}} = 4.07$$

The critical t-value is, according to Table 2-6, $t\,(1 - \alpha/2;$ $n_1 + n_2 - 2) = t\,(1 - 0.05/2;12) = 2.18$. This means that the calculated t-value is greater than the critical value, i.e. the *two-sided test* shows the result is significant. The differences between the titanium determination results from the two laboratories cannot be explained by random errors.

Testing for distributions

The previous tests served the purpose for detecting differences between means or variances. The goodness-of-fit between an observed and a hypothetical distribution is assessed by two additional tests, i.e. the γ^2- and the Kolmogorov-Smirnov's test.

γ^2-test

The γ^2 goodness-of-fit test is used to determine whether the observations of a population are sampled from a hypothetical distribution density function, e.g. the normal distribution, at a given significance level α. The null and alternative hypotheses correspond to:

H_0: the population is normally distributed with the mean $\bar{x}$ and the variance s^2

H_1: the population is not normally distributed.

To perform the test the interval for the observations is divided into k classes. The test quantity γ^2 is then obtained by comparing the theoretically expected distribution density with the observed frequency distribution according to (cf. Fig. 2-9):

$$\gamma^2 = \sum_{i=1}^{k} \frac{(h_i - f_i)^2}{f_i} \tag{2-38}$$

with h_i – observed frequency in class i
       f_i – theoretically expected frequency in class i

Fig. 2-9.
Schematic plot of the observed
frequency, h_i, and the theoretically
expected frequency, f_i, according
to a normal distribution

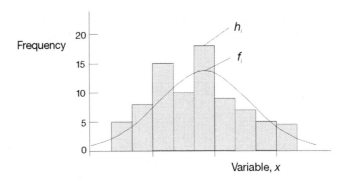

By comparison of the calculated χ^2-value with the tabulated value of the χ^2-distribution the null hypothesis is accepted if it is valid:

$$\chi^2 \leq \chi^2(1 - \alpha; k - 1 - 2)$$

and is rejected if

$$\chi^2 > \chi^2(1 - \alpha; k - 1)$$

Presuppositions for the applicability of the χ^2-test are that,
- the frequency in the middle classes is at least five and
- the frequency in the tail classes is at least one.

The test is not restricted to testing for normal distribution.

Kolmogorov-Smirnov's test for small number of samples

In practice the χ^2-test often does not work because of the above mentioned presuppositions. An alternative here is the Kolmogorov-Smirnov's test, in which a hypothetical distribution function $F_0(x)$ is used (cf. Fig. 2-7) and not the density function of the distribution, as with the χ^2-test.

In the null hypothesis the observed distribution function $F_0(x)$ is tested by:

$$H_0: F(x) = F_0(x) \tag{2-39}$$

versus the alternative hypothesis

$$H_1: F(x) \neq F_0(x) \tag{2-40}$$

For this comparison a test statistic is used to characterize the distance between the hypothetical and observed distribution functions. This distance is computed from the maximum difference between the two distribution curves as d_{max} and is compared with the critical value of the quantity $d(1 - \alpha, n)$. If $d_{max} < d(1 - \alpha, n)$ the assumed distribution is accepted. Critical d-values for testing for the *normal distribution* are given in

Table VI in the appendix. Again, in practice it is easier to evaluate the p-level (see above) provided by the common software-packages in order to decide on the significance of the test.

Example 2-7: *Kolmogorov-Smirnov's test*

The spectrophotometric measurements in Table 2-1 must be tested versus a normal distribution by means of Kolmogorov-Smirnov's test at a significance level of $\alpha = 0.05$. In the first step the empirical distribution function, $F(x)$, is evaluated as shown in Fig. 2-10. For comparison of the hypothetical distribution function the cumulative frequency obtained from the mean and the standard deviation of the data is plotted against the (standard normal) deviate z (cf. Eq. (2-28)).

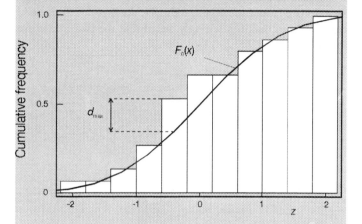

Fig. 2-10.
Determination of the test statistics for the Kolmogorov-Smirnov's test as the maximum difference d_{max} between the empirical cumulative frequency distribution of the data and the hypothetical distribution function $F_0(x)$

After evaluation of the hypothetical distribution function $F_0(x)$ the difference between both distributions is computed. The maximum difference amounts to:

$$d_{max} = 0.133$$

From Table VI in the appendix a critical value of d (0.95, 15) = 0.220 results. Since $d_{max} < d\,(1 - \alpha, n)$, it can be assumed that the photometric data belong to a normal distribution.

Errors of the first and second kind

The risk α corresponds to an error of the first kind, i.e. the null hypothesis is rejected, although it is true. The risk, however, cannot be chosen arbitrarily because errors of the second kind would then increase considerably. An error of the second kind means that the null hypothesis is accepted even though it is false (cf. Table 2-9).

Table 2-9.
Relationship between testing hypotheses and errors of the first and second kinds

Decision for	Given H_1	H_0
H_0	Correct decision $P = 1 - \alpha$	Error of the second kind $P = 1 - \beta$
H_1	Error of the first kind $P = \alpha$	Correct decision $P = 1 - \beta$

An *error of the first kind* is also termed α error, type I error or rejection error. Other names for an *error of the second kind* are β error, type II error or acceptance error.

Fig. 2-11a illustrates the relationship between an error of the first kind, also called α-error, and an error of the second kind (β-error) for comparison of two means. An error of the fisrt kind indicates that the means are taken to be different, although they deviate from each other randomly. An error of the second kind indicates that it is wrongly stated that the two means are comparable.

Fig. 2-11.
Errors of the first and second kind for larger (a) and smaller (b) standard deviations of the distribution

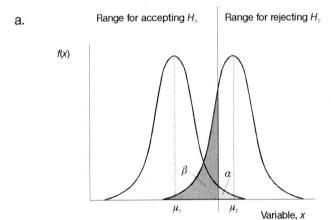

a.

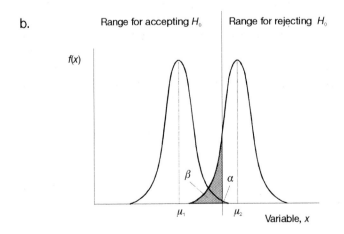

b.

A shift of the critical values to lower α-values is related to an increase of the β-error. A simultaneous decrease of the α- and β-error is only feasible if the number of measurements n is increased. Since the width of the distribution is proportional to $\sigma/\sqrt{n}$ a larger number of measurements leads obviously to a narrower distribution (cf. Fig. 2-11b).

In conclusion, the actual situation dictates the costs of the two kinds of error and indicates whether an increase in the number of measurements is advantageous.

The failure to recognise a disease, for example, is much more critical than precautionary therapy of a patient. In the latter case an error of the first kind is acceptable, i.e. from the clinical data a healthy person is diagnosed as having a disease. Failure to recognize a disease from clinically abnormal data is an error of the second kind. In this instance the data from a diseased patient lead to the diagnosis that he is healthy.

Tests for outliers

An important application of statistical tests is the recognition of outliers. Here we only consider outliers in a series of measurements. Outlier tests for methods of pattern recognition and modeling are introduced in Chapters 5 and 6, respectively.

Outliers in a series of measurements are extraordinarily small or large observations compared with the bulk of the data. Before an outlier test is applied one should determine the reason for those striking measurements. Uncritical elimination of outliers might lead to wrong conclusions. Think about the average age. A very high age of, say, 120 years would definitely rejected as an outlier with the test to be discussed below.

Dixon's Q-test

Testing for an outlier under the assumption of normal distribution can be carried out by use of Dixon's test. This test uses the range of measurements and can be applied even when only few data are available. The n measurements are arranged in ascending order. If the very small value to be tested as an outlier is denoted by x_1 and the very large value by x_n, then the test statistics are calculated by means of:

$$Q_1 = \frac{|x_2 - x_1|}{|x_n - x_1|} \quad \text{and} \tag{2-41}$$

$$Q_n = \frac{|x_n - x_{n-1}|}{|x_n - x_1|} \tag{2-42}$$

Table 2-10.
Critical values for the Q-test at the
1% risk level

n	$Q\,(0.99;\,n)$
3	0.99
4	0.89
5	0.76
6	0.70
7	0.64
8	0.59
9	0.56
10	0.53
11	0.50
12	0.48
13	0.47
14	0.45
15	0.44
20	0.39
25	0.36
30	0.34

The null hypothesis, i.e. that the considered measurement is not an outlier, is accepted if the quantity $Q < Q(1 - \alpha;\,n)$. Q-values for a selected significance level of 0.99 are given in Table 2-10.

Example 2-8: Q *Outlier test*

Trace analysis of polycyclic aromatic hydrocarbons (PAH) in a soil revealed for the trace constituent benzo[a]pyrene the following values (in mg kg^{-1} dry weight):

5.30, 5.00, 5.10, 5.20, 5.10, 6.20, 5.15

The Q-test must be used to test whether the smallest and the largest values might be outliers. The measurements arranged in ascending order are:

x_1	x_2	x_3	x_4	x_5	x_6	$x_{n=7}$
5.00	5.10	5.10	5.15	5.20	5.30	6.20

Inserting the data into Eq. (2-41) results in the following Q-values for the smallest and largest values:

$$Q_1 = \frac{|5.10 - 5.00|}{|6.20 - 5.00|} = 0.083$$

and

$$Q_n = \frac{|6.20 - 5.30|}{|6.20 - 5.00|} = 0.75$$

As critical values at level $\alpha = 0.01$ we obtain from Table 2-10 $Q(1 - \alpha = 0.99;\, n = 7) = 0.64$. For the smallest value of 0.50 we obtain $Q_1 < Q\,(1 - \alpha;\, n)$, i.e. the value cannot be marked as an outlier. For the largest value 6.20 the test indicates that $Q_n > Q\,(1 - \alpha;\, n)$, i.e. the latter value is an outlier.

Grubbs's- test

This test is also based on the assumption of a normally distributed population. It can be applied for series of measurements consisting of 3 to 150 measurements. The null hypothesis, according to which x^* is not an outlier within the measurement series of n values, is accepted at level α, if the test quantitity T is:

$$T = \frac{|\bar{x} - x^*|}{s} < T\,(1 - \alpha;\, n) \tag{2-43}$$

Where the mean $\bar{x}$ and the standard deviation s are calculated for all the values. By use of the test quantity T the distances of the suspicious values from the mean are determined and related to the standard deviation of the measurements. Critical values for the quantitity of Grubbs's-test are given in Table 2-11.

Table 2-11.
Critical values for Grubbs's-test at two significance levels

n	$T\,(0.95;\,n)$	$T\,(0.99;\,n)$
3	1.15	1.16
4	1.46	1.49
5	1.67	1.75
6	1.82	1.94
7	1.94	2.10
8	2.03	2.22
9	2.11	2.32
10	2.18	2.41
12	2.29	2.55
15	2.41	2.71
20	2.56	2.88
30	2.75	3.10
40	2.87	3.24
50	2.96	3.34

Example 2-9: *Grubbs's Outlier Test*

The data for trace analysis of benzo[a]pyrene from example 2-8 are to be investigated by the Grubbs's-test. First the mean ($\bar{x} = 5.29$) and the standard deviation ($s = 0.411$) of the data are calculated. Next the smallest and largest values are inserted into Eq. (2-43) giving:

$$T_1 = \frac{|5.29 - 5.00|}{0.411} = 0.71 \qquad T_n = \frac{|5.29 - 6.0|}{0.411} = 2.21$$

The critical test value at level $\alpha = 0.01$ is $T\,(1 - \alpha = 0.99;\ n = 7) = 2.10$. As a consequence the test result is not significant ($T_1 < T\,(1 - \alpha;\ n)$) for the smallest value but is significant ($T_n > T\,(1 - \alpha;\ n)$) for the largest. The largest value (6.20) is therefore confirmed as an outlier.

2.3 Analysis of variance

One-way analysis of variance

Analysis of variance (ANOVA) is used to analyze observations that depend on the operation of one or more effects. These effects are caused by factors the levels of which are also called groups, e.g. different laboratories.

Let us start with analysis of variance for a single factor, termed *one-way analysis of variance*. Table 2-12 demonstrates the general scheme used for the measurement of this kind of ANOVA.

Table 2-12.
Data scheme for a one-way analysis of variance

Repetition	Group			
	1	2	. . .	q
1	y_{11}	y_{12}		y_{1q}
2	y_{21}	y_{22}		y_{2q}
:				
:	$y_{n_1 1}$	$y_{n_2 2}$		$y_{n_q q}$
Mean:	$\bar{y}_1$	$\bar{y}_2$		$\bar{y}_{\text{total}}$

To test for systematic differences between the groups it is assumed that each measurement, y_{ij}, can be described by the sum of the total mean, $\bar{y}_{total}$, the group mean, y_j, and the residual random error, e_{ij}, according to:

$$y_{ij} = \bar{y}_{total} + (\bar{y}_j - \bar{y}_{total}) + e_{ij} \qquad (2\text{-}44)$$

where y_{ij} $-$ measurement of repetitions i in group j

$(\bar{y}_j - \bar{y}_{total})$ $-$ laboratory bias estimated by the group mean $\bar{y}_j$ of group j

The total variance, expressed as the sum of squares of deviations from the grand mean, is partitioned into the variances *within* the different groups and *between* the groups. This means that the sum of squares corrected for the mean, SS^2_{corr}, is obtained from the sum of squares *between* the groups, or factor levels, SS^2_{fact}, and the residual sum of squares within the groups, SS^2_R:

$$SS^2_{corr} = SS^2_{fact} + SS^2_R \qquad (2\text{-}45)$$

$$\text{with } SS^2_{corr} = \sum_{j=1}^{q} \sum_{i=1}^{n_j} (y_{ij} - \bar{y}_{total})^2 \qquad (2\text{-}46)$$

$$SS^2_{fact} = \sum_{j=1}^{q} n_j (\bar{y}_j - \bar{y}_{total})^2 \quad \text{and} \quad \bar{y}_j = \frac{1}{n_j} \sum_{i=1}^{n_j} y_{ij} \qquad (2\text{-}47)$$

$$SS^2_R = \sum_{j=1}^{q} \sum_{i=1}^{n_j} (y_{ij} - \bar{y}_j)^2 \qquad (2\text{-}48)$$

$$\bar{y}_{total} = \frac{1}{n} \sum_{j=1}^{q} \sum_{i=1}^{n_j} y_{ij} \qquad (2\text{-}49)$$

q $-$ number of groups (factor levels)
n_j $-$ number of replicate determinations per group j
n $-$ total number of measurements, i.e. $n = \sum_{j=1}^{q} n_j$

To decide on the acceptance of the null hypothesis – i.e. the groups belong to the same population and differ only randomly – an F-test is performed (cf. Eq. (2-37)):

$$F = \frac{\dfrac{SS^2_{fact}}{(q-1)}}{\dfrac{SS^2_R}{(n-q)}} \qquad (2\text{-}50)$$

The calculated F-value is subsequently compared with the critical value. If the calculated F-value is lower than the critical value, the result of the test is not significant, i.e. the groups are different only at random.

Example 2-10: *One-way analysis of variance*

The potassium concentration of water is to be determined. For this parallel determinations are carried out in four laboratories. Before aggregation of the determined values one-way analysis of variance is used to test whether there are systematic differences between the results obtained from the different labs.

Each laboratory performs triplicate determinations giving concentrations for potassium as summarized in Table 2-13 together with the lab means. The mean values obtained by the different laboratories range between 10.20 and 10.77 mg potassium per liter.

Table 2-13.
Potassium concentration in mg L^{-1} from triplicate determinations in four different laboratories

Replicates	Laboratory (group)			
	1	2	3	4
1	10.2	10.6	10.3	10.5
2	10.4	10.8	10.4	10.7
3	10.0	10.9	10.7	10.4
Mean:	10.20	10.77	10.47	10.53

The results for the sums of squares within and between the laboratories based on the potassium determinations in Table 2-13 are given in Table 2-14. This representation of data is called an ANOVA-table.

Table 2-14.
Results from one-way analysis of variance for the potassium determinations of Table 2-13

Source of variation	Degrees of freedom	Sum of squares
Between laboratories, SS^2_{fact}	3	0.489
Within laboratories, SS^2_R	8	0.260
Corrected for the mean, SS^2_{corr}	11	0.749

The F-value is calculated according to Eq. (2-50):

$$F = \frac{\dfrac{0.498}{(4-1)}}{\dfrac{0.260}{(12-4)}} = 5.02$$

The value of $F = 5.02$ is greater than the tabulated value $F(1 - \alpha = 0.95; f_1 = 3; f_2 = 8) = 4.07$ (see Table IV). The test shows therefore the result is significant, i.e. the differences between the laboratories cannot be considered to be random differences. The potassium determinations of at least one laboratory deviate systematically from the others.

Be means of one-way analysis of variance the effect of one factor can be investigated at different levels. In our example the effect of a lab on the results of determinations was tested. In many applications several factors have to be evaluated simultaneously. For example, apart from the effects of the laboratories, influences by the operator and the quality of the instrumentation are to be expected. These effects can be studied by use of two- or multi-way analysis of variance.

Two-way and multi-way analysis of variance

Consider the situation where simultaneously two effects on analytical data must be investigated. This is done by two-way analysis of variance. Given the two factors A and B the following model can be assumed:

$$y_{ij} = \bar{y}_{\text{total}} + \left(\bar{y}_i^A - y_{\text{total}}\right) + \left(\bar{y}_j^B - y_{\text{total}}\right) + e_{ij} \qquad (2\text{-}51)$$

where y_{ij} – measurement of row i and column j
$\bar{y}_i^A$ – mean of factor A in row i
$\bar{y}_j^B$ – mean of factor B in column j
e_{ij} – random error (residual)

The data are ordered such that the effects of the one factor form the rows of a matrix and those of the other factor form the columns (Table 2-15).

Table 2-15.
Data design for a two-way analysis of variance

Factor A	Factor B				Mean $\bar{y}_i$
	1	2	. . .	q	
1	y_{11}	y_{12}		y_{1q}	$\bar{y}_1^A$
2	y_{21}	y_{22}		y_{2q}	$\bar{y}_2^A$
:					
p	y_{p1}	y_{p2}		y_{pq}	$\bar{y}_p^A$
Mean $\bar{y}_i^B$	$\bar{y}_1^B$	$\bar{y}_2^B$		$\bar{y}_q^B$	$\bar{y}_{\text{total}}$

Calculation of the means is based here on the following formulae:

$$\bar{y}_i^A = \frac{1}{q}\sum_{j=1}^{q} y_{ij} \tag{2-52}$$

$$\bar{y}_j^B = \frac{1}{p}\sum_{i=1}^{p} y_{ij} \tag{2-53}$$

$$\bar{y}_{total} = \frac{1}{pq}\sum_{j=1}^{q}\sum_{i=1}^{p} y_{ij} \tag{2-54}$$

For the total sum of squares Eq. (2-45) is again valid, i.e. the total sum of squares corrected for the mean is obtained as the sums of squares of factor A, SS_A^2, factor B, SS_B^2, and a residual sum of squares, SS_R^2:

$$SS_{corr}^2 = SS_A^2 + SS_B^2 + SS_R^2 \tag{2-55}$$

with $\quad SS_{corr}^2 = \sum_{j=1}^{q}\sum_{i=1}^{p}(y_{ij} - \bar{y}_{total})^2 \tag{2-56}$

$$SS_A^2 = q\sum_{i=1}^{p}(\bar{y}_i^A - \bar{y}_{total})^2 \tag{2-57}$$

$$SS_B^2 = p\sum_{j=1}^{q}(\bar{y}_j^B - \bar{y}_{total})^2 \tag{2-58}$$

$$SS_R^2 = SS_{corr}^2 - SS_A^2 - SS_B^2 \tag{2-59}$$

In addition, measurements could be repeated for all factor combinations, as was shown for one-way analysis of variance (cf. Table 2-12). All sums must be indexed over the replicates for computation of the means (Eq. (2-52) to (2-54)).

Example 2-11: *Two-way analysis of variance*

For preparation of a standard reference sample for determination of manganese in alloyed steel a laboratory intercomparison study is carried out. Four laboratories participate and each lab uses three different analytical principles. Two-way analysis of variance is to be used to test whether there are systematic differences between the laboratories and the principles of analysis. The results of the chemical analyses are given in Table 2-16.

Table 2-16.
Analytical determinations of manganese (mass%) in steel carried out in four different laboratories on the basis of three analytical principles

Analytical principle	Laboratory				Mean $\bar{y}_i^A$
	1	2	3	4	
1	2.01	1.96	1.99	2.03	2.00
2	1.97	2.05	2.04	1.99	2.01
3	2.05	2.06	2.11	2.12	2.09
Mean $\bar{y}_i^B$	2.01	2.02	2.05	2.05	$\bar{y}_{total} = 2.03$

The result of two-way analysis of variance according to Eqs. (2-55) to (2-59) is summarized in Table 2-17. Calculation of the F-values is done separately for the two factors.

Effect of the factor *analytical principle*:

$$F_A = \frac{\dfrac{SS_A^1}{(p-1)}}{\dfrac{SS_R^1}{(p-1)(q-1)}} = \frac{\dfrac{0.00175}{3-1}}{\dfrac{0.00788}{(3-1)(4-1)}} = 6.67$$

For the critical F-value at a significance level of $\alpha = 0.05$ from Table 2-8 the value is $F\,(1 - \alpha = 0.95;\, f_1 = 2;\, f_2 = 6) = 5.14$. Since $F_A > F\,(1 - \alpha;\, f_1\,;\, f_2\,)$, the result of the test is significant, i.e. the different analytical principles lead to systematic differences.

Effect of the factor *laboratory*:

$$F_B = \frac{\dfrac{SS_B^2}{(q-1)}}{\dfrac{SS_R^2}{(p-1)(q-1)}} = \frac{\dfrac{0.00297}{4-1}}{\dfrac{0.00788}{(3-1)(4-1)}} = 0.753$$

The critical F-value at a significance level of $\alpha = 0.05$ is here $F\,(1 - \alpha = 0.95;\, f_1 = 3;\, f_2 = 6) = 4.76$. Comparison of the calculated F-value with the tabulated value shows that $F_B < F\,(1 - \alpha;\, f_1;\, f_2)$, i.e. the test result is statistically not significant. The differences between the labs are random.

In Table 2-17 besides the F-value the p-level is also given. For the significant effect of the analytical principle this value is smaller than the previously assumed level of $\alpha = 0.05$. In the case of the nonsignificant effect of the laboratory the p-level is larger than 0.05.

Table 2-17.
Two-way analysis of variance of the data in Table 2-16

Source of variation	Degrees of freedom	Sum of squares	p-level	F-value	F_{tab}-value
Principle,	2	0.01752	0.0299	6.67	5.14
Laboratory, SS_B^2	3	0.00297	0.559	0.753	4.76
Residual, SS_R^2	6	0.00788			
Corrected for the mean, SS_{corr}^2	11	0.02837			

In principle, analysis of variance is not restricted to two factors. Common software usually offers multi-way analysis where the tests can be performed for as many factors as the problem to be solved requires.

The effects of the different factors can be investigated either independently of each other or the interactions of factors can be considered in the variance analysis. The data for multi-way analysis of variance are ordered in a matrix such that the rows represent the individual runs and the columns contain the factor levels as well as the responses (Table 2-18). This treatment of data corresponds to those used in experimental designs (cf. Sec. 4.2).

Table 2-18.
Schematic representation of factors and the response for multi-way (multi-factor) analysis of variance at factor levels 1 to 4

Experiment	Factor A	Factor B	Factor C	...	Response
1	1	1	4		y_1
2	2	1	3		y_2
:					
n	4	2	3		y_n

MANOVA: multidimensional analysis of variance

The aforementioned treatment was limited to the study of factors on a single response or measurement, i.e. only *one* dependent variable was investigated. In our examples this was the potassium concentration in water or the manganese content of steel. More generally it can be of interest to estimate the effects of factors on a complete spectrum or on an elemental pattern. For this purpose an analysis of variance is needed that can handle several responses, i.e. a multi-dimensional analysis of variance or MANOVA.

The data to be studied are arranged by analogy with the design in Table 2-18, with the difference, however, that several columns for the different dependent variables, e.g. the wavelength of an optical spectrum, become necessary.

As the test statistics of a MANOVA the Wilks λ-value can be applied. This value is defined by the i eigenvalues of the data matrix λ_i as follows:

$$\lambda = \prod_i \left(\frac{\lambda_i}{1 + \lambda_i} \right) \qquad (2\text{-}60)$$

For Wilks λ-values close to 0 the null hypothesis is rejected according to which the effect of the factors can be attributed to random effects. For λ-values close to 1 no factor effect can be deduced. The significance of the test results is again to be derived from testing the Wilks λ-value against a theoretical distribution. At least for large n this distribution is the χ^2-distribution. In practice the p-level should be considered as given by common software packages and compared with the risk level α.

2.4 General reading

Anderson, R. L., *Practical Statistics for Analytical Chemistry*: Van Nostrand Reinhold, New York, 1987.

Dunn, O. J., Clark, V. A., *Applied Statistics: Analysis of Variance and Regression*: Wiley, New York, 1974.

Graham, R. C., *Data Analysis for the Chemical Sciences – A Guide to Statistical Techniques*: VCH, New York, 1993.

Miller, J. C., Miller, J. N., *Statistics for Analytical Chemistry*: Second Edition, Ellis Horwood, Chichester, 1988.

Quantifying Uncertainty in Analytical Measurements, EURACHEM, London, 1995.

Questions and problems

1. Calculate the following descriptive statistics for the data on water hardness (mmol L^{-1}) given below: arithmetic mean, median, standard deviation, variance, standard error, confidence interval at a significance level of 0.01, range, and the interquartile distance.

   8.02; 7.84; 7.98; 7.95; 8.01; 8.07; 7.89

2. Characterize the data in problem 1 graphically by means of a box-whisker plot.

3. Determination of vitamin E in salad-oil was carried out by a routine voltammetric method and by flow injection analysis (FIA). The following results (%mass) were obtained:

Sample	Photometric	FIA
1	32.1	31.9
2	32.3	31.8
3	31.9	31.7
4	32.1	31.8
5	32.0	31.6
6	32.1	31.9
7	31.8	31.8

Compare the precision of the two methods by means of an *F*-test at a significance level of $\alpha = 0.05$.

4. Decide on the basis of the result in problem 3 by use of an appropriate *t*-test whether the different results of the two methods are significant at a level of $\alpha = 0.05$.

5. Use Dixon's and Grubbs' methods to test for outliers in the water hardness data in problem 1.

6. Five charges of gasoline (A, B, C, D, E) must be compared with respect to their octane rating. To account for confounded effects from the analyst and the day of analysis a 5×5 Latin square is used in the experimental design.

Day	Analyst				
	1	2	3	4	5
1	A	B	D	C	E
2	B	C	D	E	A
3	C	D	E	A	B
4	D	E	A	B	C
5	E	A	B	C	D

The octane ratings obtained were:

Day	Analyst				
	1	2	3	4	5
1	96.6	96.6	95.7	96.1	97.0
2	96.2	96.2	95.3	96.5	95.9
3	96.2	96.0	95.8	95.8	95.8
4	94.9	96.6	95.9	96.3	95.7
5	96.0	96.2	96.1	95.2	95.9

Use a commercial software package enabling multifactor ANOVA to decide whether there are systematic differences between the five charges at a significance level of $\alpha = 0.05$.

3 Signal processing and time-series analysis

Learning objectives

- To learn about digital filters, such as the nonrecursive moving-average and polynomial filters as well as the recursive Kalman filter
- To apply filters for smoothing, derivation and integration of signals
- To introduce signal manipulation based on Fourier and Hadamard transformations and their appropriate usage for filtering, convolution, deconvolution, integration, data reduction and background correction
- To understand interpolating and smoothing splines
- To characterize correlated data by autocorrelation and cross-correlation functions

In this chapter we first learn about methods that can be used for *signal processing*. Analytical signals are recorded as spectra, chromatograms, voltammograms or titration curves. These signals are monitored in the frequency, wavelength or time domains. Typical aims are smoothing and filtering, derivation, area determination, transformation or deconvolution of signals.

Secondly we investigate *time series*, e.g., pH-values measured in a lake over one year. In contrast to the independent random data considered in Chapter 2 here we deal with correlated data. The observations are assumed to be realizations of a stochastic process where the observations made at different time-points are statistically dependent. In order to recognize drifts, periodicities or noise components in a time series, correlation within the time series must be investigated.

Methods for evaluation of analytical signals are: transformation, smoothing, correlation, convolution, deconvolution, derivation and integration.

3.1 Signal processing

Digital smoothing and filtering

The extensive use of computers in the analytical laboratory is responsible for the fact that data are usually processed *digitally* (cf. Sec. 1.1). As a consequence further processing of the digital signals is carried out by software rather than by hardware as performed hitherto.

Therefore, digital filters dominate analog filters. Digital filters have the following advantages compared with the analog variety:

- The digitized data can immediately be processed further.
- Temperature- or time-dependent noise does not exist. The only noise components to be accounted for are round-off errors.
- Digital filters are easier to modify and usually easier to understand.
- The filters implemented as software can be transferred to other computers.
- The filter can be easily optimized and tailor-made to specific processes.

Preprocessing of analytical signals by filtering serves the following purpose:

- Enhancement of the signal versus the noise.
- Derivation of signals for subtraction of background and for improvement of visual resolution.
- Integration for quantitative signal evaluation.

Moving-average filter

Averaging is one means of smoothing of data.

The simplest filter operates on the basis of moving averages (Fig. 3-1). Within a predefined window a weighted linear combination of all the signal values is produced. The window determines the *filter width*. The window is moved successively along the equally spaced data. For calculation of the filtered value, y_k^*, from the raw signal values, y_k, the following relationship is used:

$$y_k^* = \frac{1}{2m+1} \sum_{j=-m}^{j=m} y_{k+j} \tag{3-1}$$

where k is the index for the actual data point, $2m + 1$ represents the size of the window (filter width), and m is the variable for adjusting the filter width.

Fig. 3-1.
Moving-average filter for a filter-width of $2m + 1 = 3$, i.e. $m = 1$. Note that for the extreme points no filtered data can be calculated, since they are needed for computing the first and last average. ● original signal value; o filtered signal value

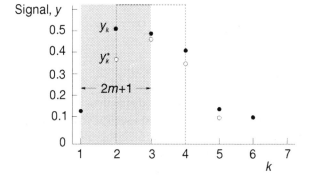

After applying the moving-average filter the data contain *less noise*. With structured data the filter-width must be chosen such that the structure of the data, e.g., of a peak, is not distorted.

Fig. 3-2 demonstrates the filtering of raw data by use of a five-point moving-average filter (curve 1). In this example a filter-width of five points begins to lead to distortion of the peaks. This effect is enhanced if the filter-width is further increased as demonstrated here for an 11-point filter (Fig. 3-2, curve 2). The appropriate choice of filter-width will be discussed below.

Too large a filter-width reduces the original height of a peak and leads simultaneously to its broadening.

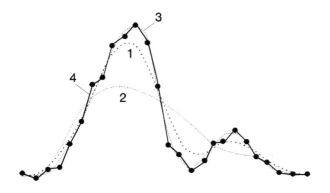

Fig. 3-2.
Filtering of a discrete analytical signal consisting of k data points with different filters:
(1) five-point moving-average filter, (2) 11-point moving-average filter, (3) five -point Savitzky-Golay filter, (4) interpolated signal

Polynomial smoothing: Savitzky-Golay filter

Very efficient smoothing of data is obtained with filters that weight the raw data differently. With the moving-average filter all the data were weighted by the same factor, i.e. $1/(2m + 1)$ (cf. Eq. 3-1). A better fit results if weights are used that approximate the data by means of a polynomial of higher order (Savitzky-Golay filter). The filter coefficients are identical for second and third-order polynomials.

After deciding on the filter-width the filtered value for the kth data point is calculated from:

$$y_k^* = \frac{1}{NORM} \sum_{j=-m}^{j=m} c_j y_{k+j} \qquad (3\text{-}2)$$

where *NORM* is a normalization factor obtained from the sum of the coefficients c_j.

The filter coefficients c_j are tabulated in Table 3-1 for different filter-widths. Fig. 3-2, curve 3, demonstrates the effect of a Savitzky-Golay filter with a filter-width of five points applied to the raw data. Compared with the five-point moving-average filter the better fit is clearly apparent.

Table 3-1.

Coefficients of the Savitzky-Golay filter for smoothing based on a quadratic/cubic polynomial according to Eq. 3-2

Points	25	23	21	19	17	15	13	11	9	7	5
−12	−253										
−11	−138	−42									
−10	−33	−21	−171								
−9	62	−2	−76	−136							
−8	147	15	9	−51	−21						
−7	222	30	84	24	−6	−78					
−6	287	43	149	89	7	−13	−11				
−5	322	54	204	144	18	42	0	−36			
−4	387	63	249	189	27	87	9	9	−21		
−3	422	70	284	224	34	122	16	44	14	−2	
−2	447	75	309	249	39	147	21	69	39	3	−3
−1	462	78	324	264	42	162	24	84	54	6	12
0	467	79	329	269	43	167	25	89	59	7	17
+1	462	78	324	264	42	162	24	84	54	6	12
+2	447	75	309	249	39	147	21	69	39	3	−3
+3	422	70	284	224	34	122	16	44	14	−2	
+4	387	63	249	189	27	87	9	9	−21		
+5	322	54	204	144	18	42	0	−36			
+6	287	43	149	89	7	−13	−11				
+7	222	30	84	24	−6	−78					
+8	147	15	9	−51	−21						
+9	62	−2	−76	−136							
+10	−33	−21	−171								
+11	−138	−42									
+12	−253										
NORM	5175	8059	3059	2261	323	1105	143	429	231	21	35

Filter selection

For selecting the most appropriate filter some rules are followed:

- If the filter is applied to the data *repetitively* the largest smoothing effect (>95%) is observed in the first application. Therefore, single smoothing is usually sufficient.
- The *filter-width* should correspond to the full-width-at-half-maximum of a band or a peak. Too small a filter-width results in unsatisfactory smoothing. Too large a filter-width leads to distortion of the original data structure (cf. Fig. 3-2).
- *Distortion* of data structure is more severe in respect of the area than of the height of the peaks. Therefore, the filter-width selected must be smaller if the height rather than the area is evaluated.

The influence of the filter-width on the distortion of peaks can be quantified by means of the *relative filter-width*, $b_{relative}$, according to:

$$b_{relative} = \frac{b_{filter}}{b_{0.5}} \qquad (3\text{-}3)$$

where b_{filter} is the filter-width, and $b_{0.5}$ is the full-width-at-half-maximum.

To illustrate the influence of filter-width on the distortion of peak shape the error of evaluating the height and area of Gaussian and Lorentz peaks by polynomial smoothing is shown in Fig. 3-3 (cf. Fig. 3-4).

If the peak area is measured distortions become important only at relative filter-widths greater than 1. In contrast, when measuring peak height the relative filter-width selected should not be much larger than 0.5.

From Fig. 3-3 it also can be understood that the distortion of peaks is dependent on their *shape*. Lorentz peaks are less distorted than Gaussian peaks.

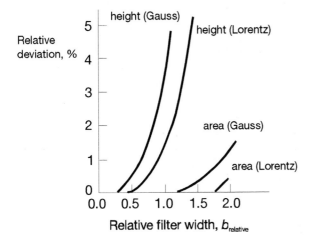

Fig. 3-3.
Dependence of the relative error for smoothing of Gaussian and Lorentz peaks on the relative filter-width (Eq. 3-3)

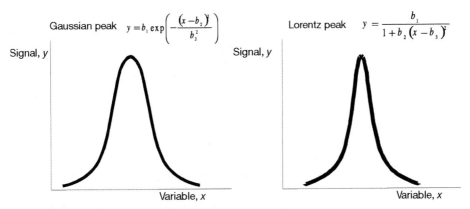

Gaussian peak $\quad y = b_1 \exp\left(-\frac{(x - b_2)^2}{b_3^2}\right)$ Lorentz peak $\quad y = \frac{b_1}{1 + b_2\,(x - b_3)^2}$

Fig. 3-4. Shapes of Gaussian (cf. Eq. 2-2) and Lorentz peaks. b_1, b_2 and b_3 are constants

Example 3-1: *Savitzky-Golay filter*

The following data are to be smoothed by a quadratic five-point polynomial filter:

Data point k:	1	2	3	4	5	6	7	8	9	10	11
Signal value y_k:	0.11	0.52	0.49	0.41	0.30	0.27	0.16	0.15	0.12	0.08	0.02

By using the coefficients from Table 3-1 we obtain for the smoothing polynomial of the kth data point:

$$y_k^* = -\frac{3}{35}y_{k-2} + \frac{12}{35}y_{k-1} + \frac{17}{35}y_k + \frac{12}{35}y_{k+1} - \frac{3}{35}y_{k+2}$$

As the first smoothed value the third data point can be calculated, i.e.:

$$y_3^* = -\frac{3}{35}y_{3-2} + \frac{12}{35}y_{3-1} + \frac{17}{35}y_3 + \frac{12}{35}y_{3+1} - \frac{3}{35}y_{3+2}$$

$$= -\frac{3}{35}0.11 + \frac{12}{35}0.52 + \frac{17}{35}0.49 + \frac{12}{35}0.41 - \frac{3}{35}0.30 = 0.366$$

After moving the filter by one data point to the right a value of 0.398 is obtained for the fourth data point, and so on.

Recursive filter: Kalman filter

The above mentioned filters use raw data only and operate non-recursively. *Recursive filters* use smoothed data also. They were developed for smoothing fast processes in real time. The most popular recursive filter is the *Kalman filter*, which was developed to control the height of a missile skimming across the waves over sea and land.

A Kalman filter is based on a *dynamic system model*

$$x(k) = Fx(k-1) + w(k-1) \tag{3-4}$$

and the *measurement model*:

$$y(k) = H^T(k)\,x(k) + v(k) \tag{3-5}$$

where x represents the state vector, y is the measurement, F is the system transition matrix and H represents the measurement vector (or matrix). System noise is characterized by the vector w and measurement noise by vector v. The index k again denotes the actual measurement or time.

The recursive algorithm for Kalman filtering operates according to the scheme in Table 3-2.

Table 3-2.
Kalman filter algorithm

Propagation of the filter states with time

$$x(k) = F(k) x(k-1) \qquad (3\text{-}6)$$

Propagation of state covariance with time

$$P(k \mid k-1) = F(k) P(k-1 \mid k-1) F^{\mathrm{T}}(k) + Q(k) \qquad (3\text{-}7)$$

Kalman gain

$$\begin{aligned} K(k) &= P(k \mid k-1) H(k) \\ &\times \left[H^{\mathrm{T}}(k) P(k \mid k-1) P(k \mid k-1) H(k) + R(k) \right]^{-1} \end{aligned} \qquad (3\text{-}8)$$

State estimate update

$$x(k \mid k) = x(k \mid k-1) + K(k) \left[y(k) - H^{\mathrm{T}}(k) x(k \mid k-1) \right] \qquad (3\text{-}9)$$

Error covariance update

$$P(k \mid k) = P(k \mid k-1) - K(k) H^{\mathrm{T}}(k) P(k \mid k-1) \qquad (3\text{-}10)$$

For the system noise Q and the measurement noise R the following assumptions are valid:

$$Q(k) = E \left[w(k) w^{\mathrm{T}}(k) \right] \qquad (3\text{-}11)$$

$$R(k) = E \left[v(k) v^{T}(k) \right] \qquad (3\text{-}12)$$

Sequential estimation of the filter parameters requires initial values to be set for x and the covariance matrix P. The latter quantity is chosen in the sense of a measure of the uncertainty in the initial estimate of x. With correct models, the sequential estimates $x(k)$ quickly become independent of the initial guess $x(0)$ provided sufficiently large values are chosen for $P(0)$, e.g. 1 to 100 times $x(0)$ work well for the diagonal values of $P(0)$, the off-diagonal elements may be set to zero. A system error is to be specified only if the matrix F is not the identity matrix.

The Kalman filter can be applied for filtering, smoothing and prediction. Additional applications are known for corrections of drift and in multicomponent analysis.

Signal derivatives

The Savitzky-Golay filter can also be used for derivation of signal curves. For this appropriate filter coefficients must be inserted; these are given in the appendix – in Table VII for the first derivative and in Table VIII for the second derivative.

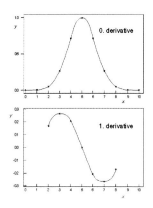

0. derivative

1. derivative

Derivative of a peak calculated
by means of the Savitzky-Golay
coefficients for the first derivative.

For example, the first derivative, y'_k, is obtained on the
basis of a five-point quadratic polynomial filter from:

$$y'_k = -\frac{2}{10}y_{k-2} - \frac{1}{10}y_{k-1} + \frac{0}{10}y_k + \frac{1}{10}y_{k+1} + \frac{2}{10}y_{k+2}$$

Derivatives are useful for eliminating *background* of a signal, for determining the *peak position* and for improving the
visual resolution of peaks.

In Fig. 3-5 the second derivative of a peak is shown. At the
peak maximum the derivative has a pronounced minimum
which is suitable for evaluation of the peak position. Compared with the original signal the full-width-at-half-maximum

Fig. 3-5.
Second derivative of a peak based
on a Lorentz function

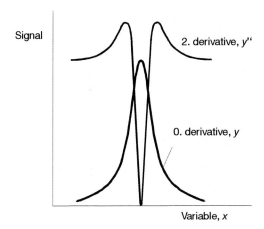

Fig. 3-6.
Visual resolution of two Lorentz
peaks with a resolution of 0.4
full-width–at-half-maximum
(a) after formation of the second
derivative (b).

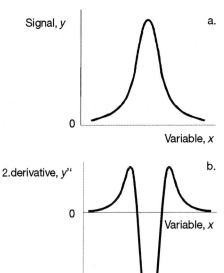

The second derivative of a peak is
easy to interpret, because it has a
'*negative peak*' shape.

is smaller for the peak in the second derivative. As a consequence two peaks can be distinguished in the second derivative even though they are not recognized in the original signal. In Fig. 3-6 this is demonstrated for two peaks with a distance of 0.4 units of full-width–at-half-maximum.

Note that the noise increases if a signal derivative is obtained. This is especially important for quantification on the basis of derivatized signals.

Example 3-2: *Noise characteristics for derivatives of signals*

The following y-signal values are obtained around an observation point k:

y_{k-2}	y_{k-1}	y_k	y_{k+1}	y_{k+2}
0.2	0.5	0.7	0.4	0.1

At this point k the filtered value is to be calculated up to the second derivative on the basis of a five-point Savitzky-Golay filter.

To characterize the corresponding noise we consider the error propagation for the polynomial filter. For the Savitzky-Golay filter (cf. Eq. 3-2) the result of error propagation (cf. Table 2-4) is expressed here by the standard deviation of the smoothed signal at point k, s_{y_k*}:

$$s_{y_k*} = \sqrt{\frac{1}{NORM^2} \sum_{j=-m}^{j=m} c_j^2 y_{k+j}}$$

(3-13)

Calculation of the filtered values and their standard deviations by means of the tabulated filter coefficients leads to:

$$y_k^* = \frac{1}{35}((-3)\cdot 0.2 + 12\cdot 0.5 + 17\cdot 0.7 + 12\cdot 0.4 + (-3)\cdot 0.1)$$
$$= 0.623$$

Zeroth derivative:

$$s_{y_k^*} = \frac{1}{35}\left((-3)^2\, 0.2 + 12^2\, 0.5 + 17^2\, 0.7 + 12^2\, 0.4 + (-3)^2\, 0.1\right)^{1/2}$$
$$= 0.522$$

relative error: $\dfrac{s_{y_k*}}{|y_k^*|} = \dfrac{0.522}{0.623} = 0.838$

First derivative:

$$y'_k = \frac{1}{10}((-2)\cdot 0.2 + (-1)\cdot 0.5 + 0\cdot 0.7 + 1\cdot 0.4 + 2\cdot 0.1)$$
$$= -0.03$$

$$s_{y'_k} = \left(\frac{1}{10^2}\left((-2)^2\, 0.2 + (-1)^2\, 0.5 + 0^2 0.7 + 1^2 0.4 + 2^2 0.1\right)\right)^{1/2}$$
$$= 0.145$$

relative error: $\dfrac{s_{y'_k}}{|y'_k|} = \dfrac{0.145}{0.03} = (4.83)$

Second derivative:

$$s_{y''_k} = \left(\frac{1}{7^2}\left(2^2 0.2 + (-1)^2\, 0.5\right.\right.$$
$$\left.\left. + (-2)^2\, 0.7 + (-1)^2\, 0.4 + 2^1 0.1\right)\right)^{1/2} = 0.316$$

$$y''_k = \frac{1}{7}(2\cdot 0.2 - 1\cdot 0.5 - 2\cdot 0.7 - 1\cdot 0.4 + 2\cdot 0.1) = -0.243$$

relative error: $\dfrac{s_{y''_k}}{|y''|_k} = \dfrac{0.316}{0.243} = 1.301$

Comparion of the relative error between the zeroth and second derivative reveals an increase by a factor of 1.5. The error propagation for the first derivative demonstrates the limits of the procedure. Since the first derivative is close to zero at the peak maximum, an unrealistically large value is obtained for the relative error.

Integration for area determination

The area under a chromatographic peak, a spectroscopic band or in thermal analysis is often directly proportional to the analyte concentration. Integration of signals is, therefore, important for quantitative data evaluation.

In the simplest case the area is calculated from the sum of the signals obtained over the corresponding variable, such as time, energy or wavelength. Because it is known, however, that random variations at the edge of a peak are much more severe than at its middle, the integration formula should give more weight to the middle signals than to those at the edges. This leads to integration formulas that are known as the trapezoidal rule or Simpson rule.

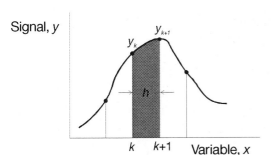

Signal, y

y_{k+1}

y_k

h

k $\quad$ $k+1$ $\qquad$ Variable, x

Fig. 3-7.

Trapezoidal rule for integration
of a signal

Fig. 3-7 demonstrates the trapezoidal rule. The area, A_T, is calculated from the sum of the signal intensities, y_k, times the distance between two successive data points at the abscissa, h, by use of:

$$A_T = \int_X y dx = h\left[\frac{1}{2}y_1 + \frac{1}{2}y_n + \sum_{k=2}^{n-1} y_k\right] \qquad (3\text{-}14)$$

Note that the edge points ($k = 1$ and $k = n$) are counted only once and so are multiplied by a factor of 1/2 in Eq. 3-14.

Still better results are obtained by the Simpson rule. This is based on approximation by a quadratic polynomial. Here the area, A_S, is obtained from:

$$A_S = \int_X y dx = \frac{h}{3}[y_1 + 4y_2 + 2y_3 + \ldots + 4y_{n-1} + y_n] \qquad (3\text{-}15)$$

In general the Simpson rule reads:

$$A_S = \frac{h}{3}\left[y_1 + y_n + 4\sum_{k=2}^{n-1} {}_I y_k + 2\sum_{k=3}^{n-2} {}_{II} y_k\right] \qquad (3\text{-}16)$$

Where the sum Σ_I is to be taken over all the even numbers and the sum Σ_{II} over all the odd numbers.

A prerequisite for application of the Simpson rule according to Eq. 3-16 is an odd number n of data points, i.e. an even number of equidistant intervals. When the number of data intervals is odd integration is started on the basis of the Simpson rule over an even number of intervals and the remaining area is calculated by another method. Typically the 3/8-rule, based on a cubic polynomial, is used as the second method.

Suppose for an odd number of intervals the signal values are to be integrated from $k = 1$ to $k = n - 3$ by the Simpson rule. Then the remaining area is obtained by the 3/8-rule according to:

$$A_{3/8} = \frac{3}{8}[y_{n-3} + 3y_{n-2} + 3y_{n-1} + y_n] \qquad (3\text{-}17)$$

61

In that way area determinations can be performed for an arbitrary number of data points.

Example 3-3: *Integration by means of the Simpson rule*

For the signal values in Example 3-1 the area is to be determined on the basis of a quadratic polynomial. According to Eq. 3-16 the resulting area is:

$$A_S = \frac{1}{3}[0.11 + 0.02 + 4(0.52 + 0.41 + 0.27 + 0.15 + 0.08)$$
$$+ 2(0.49 + 0.30 + 0.16 + 0.12)] = 2.66$$

Transformations: Fourier and Hadamard

Mathematical transformations of raw data can be used for filtering if the transformed data are multiplied by appropriate filter functions and are subsequently back-transformed into the original data domain. Most frequently the Fourier transformation (FT) is applied. In addition we also will learn about Hadamard transformation (HT).

Apart from the filtering of data, transformations are also useful for convolution and deconvolution of analytical signals, for integration, for background correction and for reducing the number of data points, e.g. in a spectrum.

Fourier transformation

A signal in the time domain can be represented by a combination of periodic functions. Fig. 3-8 demonstrates this for the combination of two sine functions.

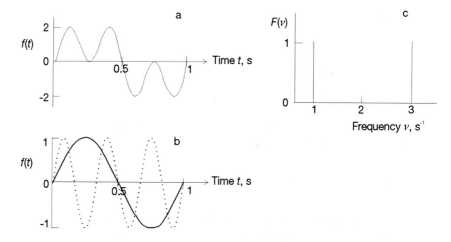

Fig. 3-8. Fourier transformation: the sum signal (a) contains the sine functions with the periods $t_1 = 1s$ and $t_2 = 1/3$ s (b). The dependence of intensity on frequency after Fourier transformation is given in (c)

Transformation of the signal from the time domain, $f(t)$, into the frequency domain, $F(\nu)$, provides two frequencies at $1\ \text{s}^{-1}$ and $3\ \text{s}^{-1}$. Because any time-dependent or, in general, continuous signal can be considered a combination of sine and cosine functions, the Fourier transformation is widely applicable.

In the time domain information is obtained on the total amplitude, but not on the frequencies of the signal. The frequency domain contains information on the frequencies and on the corresponding amplitudes of the signal, but information on the total amplitude is lost.

In conventional spectroscopy measurements are obtained in the frequency domain: the intensity of radiation is recorded as a dependence on the frequency or on the reciprocal wavelength. Some analytical methods, such as FTIR or pulsed NMR spectroscopy, provide the information in the time domain. There the opposite transformation into the frequency domain is of interest.

The maximum frequency that can be observed depends on the spacing of Δt according to:

$$\nu_{max} = \frac{1}{2\Delta t}$$

This is called the *Nyquist frequency*.

Discrete Fourier transformation

Digitized signal values in the time domain can be directly treated by discrete FT. For n discrete, equally spaced signal values we obtain for the transformation into the frequency domain:

Aliased frequencies are different frequencies which exist in a signal at the same time.

$$F(\nu) = \frac{1}{n}\sum_{t=1}^{n} f(t)e^{-(j2\pi\nu t/n)} \qquad (3\text{-}18)$$

where $j = \sqrt{-1}$

The results are complex numbers consisting of real and imaginary parts. The first term, $F(1)$, is always real and corresponds to the average of the data.

The following equation is used to express the exponential function:

$$e^{-[j2\pi\nu/n]} = \cos(2\pi\nu t/n) - j\sin(2\pi\nu t/n) \qquad (3\text{-}19)$$

Inverse Fourier transformation

Back-transformation from the frequency domain into the time domain is carried out by inverse FT, i.e.:

$$f(t) = \sum_{\nu=1}^{n} F(\nu)\,e^{(j2\pi\nu t/n)} \qquad (3\text{-}20)$$

In practice an algorithm for fast FT of data is applied. There the number n of the k data points must be a power of two, i.e. $n = 2^k$. If complex conjugated numbers are used for Eq. 3-20, forward and backward transformation can be performed with the same algorithm.

Hadamard transformation

Wavelet transformations enable the use of arbitrary base functions.

As an alternative to Fourier transformations the Hadamard transformation (HT) can be applied. The latter differs from FT in the base function. HT is based on the Walsh function in contrast to the sine and cosine functions in FT (cf. Eq. 3-19). Fig. 3-9 demsonstrates the Walsh function with boundaries of ±1.

Fig. 3-9.
The Walsh function as the base function of the Hadamard transformation

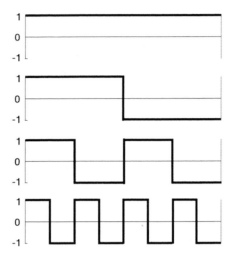

Transformation of the original data, y, into the transformed values, y^*, can be easily represented by the equation:

$$y^* = H\,y \tag{3-21}$$

where H is the $(n \times n)$ Hadamard transformation matrix, y is the vector of the original n signal values, and y^* is the vector of the transformed n signal values.

As for the fast Fourier transformation (FFT) (cf. Eq. 3-20) the total n of the k data points that are to be used must be a power of two, i.e. $n = 2^k$. The kth Hadamard transformation matrix is obtained by a simple iteration rule:

$$H_k = \begin{pmatrix} H_{k-1} & H_{k-1} \\ H_{k-1} & -H_{k-1} \end{pmatrix} \tag{3-22}$$

Example 3-4: *Hadamard transformation*

Four data points are to be treated by the HT. With $n = 2^k = 4$ we have $k = 2$. If we set $H_0 = 1$ then we obtain for the matrices H_1 and H_2 iteratively:

$$H_1 = \begin{pmatrix} 1 & 1 \\ 1 & -1 \end{pmatrix}$$

$$H_2 = \begin{pmatrix} H_1 & H_1 \\ H_1 & -H_1 \end{pmatrix} = \begin{pmatrix} 1 & 1 & 1 & 1 \\ 1 & -1 & 1 & -1 \\ 1 & 1 & -1 & -1 \\ 1 & -1 & -1 & 1 \end{pmatrix}$$

The transformation equation is, according to Eq. 3-21:

$$\begin{pmatrix} y_1^* \\ y_2^* \\ y_3^* \\ y_4^* \end{pmatrix} = \begin{pmatrix} 1 & 1 & 1 & 1 \\ 1 & -1 & 1 & -1 \\ 1 & 1 & -1 & -1 \\ 1 & -1 & -1 & 1 \end{pmatrix} \begin{pmatrix} y_1 \\ y_2 \\ y_3 \\ y_4 \end{pmatrix}$$

Multiplication of the equations provides the transformed signal values:

$$y_1^* = y_1 + y_2 + y_3 + y_4 \tag{3-23}$$

$$y_2^* = y_1 + y_2 + y_3 + y_4 \qquad \text{and so on.}$$

Insert your own numbers to transform your signals.

HT requires only simple arithmetic operations in the form of addition and subtraction, in contrast with FT calculations where complex numbers and trigonometric functions have to be processed. As a consequence the algorithm for fast Hadamard transformation (FHT) is faster by a factor of approximately three compared with the FFT algorithm.

An additional advantage is that the result of a Hadamard transformation is real, i.e. there is no imaginary part.

Application of FT and HT

In many applications of filtering, integration or data reduction the Fourier and Hadamard transformations provide comparable results. If possible, both transformations should be tested in the actual application.

A particular transformation might be advantageous if one of its specific properties is required. Thus HT is favored over FT if the speed of computation is important to the transformation. Furthermore, flat graphs, e.g. an empty spectral range, is better approximated by HT than by FT, because FT models flat segments with waves of the highest possible frequency, which are poor substitutes for a straight line. On the other hand, FT is better suited to the description of rounded shapes, such as structured bands or peaks. For these HT leads to undesired signal characterization because of its 'box'-wave-like base function.

In the following text selected applications of both transformations will be outlined.

Signal filtering

For FT filtering the signals are transformed from the time domain into the frequency domain, $F(\nu)$, by use of Eq. 3-18. After that, multiplication by a filter function, $H(\nu)$, and back-transformation with the inverse FT function (Eq. 3-20) is performed. The filtered data, $G(\nu)$, are obtained from:

$$G(\nu) = F(\nu)\,H(\nu) \qquad (3\text{-}24)$$

Since the signal of interest lies usually in the intermediate frequency range, the filter function serves the purpose of filtering very high or low frequencies. To suppress high frequencies that are typical of noise, a *low-pass filter* is required. For filtering low frequencies, that correspond, for example, to drift, a *high-pass filter* must be used (Fig. 3-10).

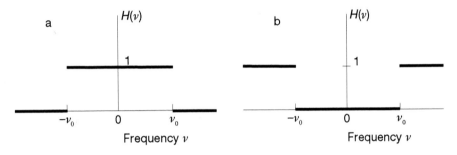

Fig. 3-10. Low-pass filter (a) and high-pass filter (b) in the frequency domain

Fourier transformations are especially suitable if different frequency parts are present in the original signal or if, for example, the ac mains frequency must be removed. Important information about the type of noise can be derived from spectral analysis in the frequency domain and as a consequence the signal-to-noise ratio can be systematically improved. *Shot noise* is recognized by a uniform frequency spectrum. This is typical of thermal noise or quantization noise from photomultiplier tubes. A continuously changing frequency spectrum is observed in the case of drift *(1/f-noise)*. *Interference noise* is characterized by a specific frequency (range), e.g. by a superposed 50 Hz power frequency.

Convolution and deconvolution

FT can be used to advantage for restoration of an analytical signal distorted by an instrument function, or for deconvolution of overlaping signals (cf. Fig. 3-11).

Let us denote the dependence of undistorted signal on time by $f(t)$, the overlapped function, e.g. instrument or Gaussian function, by $h(t)$ and the observed function by $g(t)$. Convolution (denoted by *) of the original signal with the interfering function is expressed by:

$$g(t) = f(t) * h(t) \tag{3-25}$$

In the frequency domain this corresponds to a simple scalar product of the Fourier-transformed functions:

$$G(v) = F(v)\,H(v) \tag{3-26}$$

In order to evaluate the undistorted signal, $f(t)$, Eq. 3-26 must be solved for $F(v)$, i.e.

$$F(v) = \frac{G(v)}{H(v)} \tag{3-27}$$

The result is back-transformed into the time domain according to Eq. 3-20.

Components of an analytical signal in different frequency ranges:

Frequency	Component
intermediate	peak, band
low	drift
high	noise

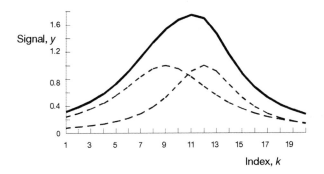

Fig. 3-11.
Decomposition (deconvolution) of peaks into two underlying individual peaks by means of FT

Note the similarity to signal filtering in Eq. 3-24. One difference from deconvolution is the type of interfering function $H(v)$. In filtering this function is a simple step function; for deconvolution of signals it must be tested whether a trapezoidal, triangular, Bessel, cosine or Gaussian function would be best.

Problems arise in the case of noise. If the signal noise is denoted by $N(v)$ Eq. 3-24 is modified to:

$$G(v) = F(v)\,H(v) + N(v) \tag{3-28}$$

This leads to a different deconvolution in the frequency domain, i.e.:

$$\frac{G(v)}{H(v)} = F(v) + \frac{N(v)}{H(v)} \tag{3-29}$$

and

$$\hat{F}(v) = F(v) + \frac{N(v)}{H(v)} \tag{3-30}$$

67

The function estimate obtained after deconvolution, $\hat{F}(v)$, does not correspond to the FT of the undistorted signal, but also contains, in the frequency domain, an undesired noise component.

An additional problem in FT deconvolution results from the finite number of data points. Back-transformation frequently leads to wave-like curves that cannot be attributed to real periodicities of the Fourier transform. To suppress those undesired side-effects an *apodization function* in form of a triangular or parabolic function is applied. By analogy with Eq. 3-27 the deconvoluted signal is calculated by using an apodization function, $D(v)$:

$$F(v) = \frac{D(v)\,G(v)}{H(v)} \tag{3-31}$$

For deconvolution of signals the specific form of the interfering signal must be determined separately. So, for example, for deconvolution of an overlaping chromatographic signal into Gaussian peaks (cf. Eq. 2-2) the full-width-at-half-maximum or the standard deviation must be known.

Integration

From the equations in Example 3-4 it can be recognized that the first row in the HT-matrix, that leads to the first Hadamard coefficient, y_1^*, by multiplication with the original data, is equal to the sum of all signal values (Eq. 3-23). On the basis of that sum the integral over all signal values can be deduced, if, e.g., the trapezoidal formula according to Eq. 3-14 is applied. The area, A, is calculated by subtraction of half of the sums of the first and last signal values:

$$A = y_1^* - \frac{1}{2}(y_1 + y_n) \tag{3-32}$$

A similar result is obtained for FT. According to Eq. 3-18 the first Fourier coefficent, $F(1)$, corresponds to the average of the signal values. Multiplication with the data number n provides the sum over all values for integration by Eq. 3-32. In the latter case we assumed that only the real part of FT is considered. Fortunately, for any real function the first coefficient, $F(1)$, should be always real.

Data reduction and background correction

Reduction of data points is important if, e.g., further processing of a spectrum is feasible only if the number of data points is reduced. For reduction of measurements in the original data vector the data are transformed by means of FT or HT. Afterwards back-transformation is performed on the basis of a limited number of Fourier or Hadamard coefficients. For back-transformation the coefficients are sorted according to importance and the effect of less important coefficients is thus eliminated (cf. Zupan, Sec. 3.3). In practice the number of coefficients is not changed, but unimportant coefficients are set to zero.

If the most significant coefficient $F(2)$ for the FT or y_2^* for the HT is set to zero, then a background correction results because the coefficients correspond to the base functions of those transformations. Because of the 'box'-wave-like base function of the HT (cf. Fig. 3-9) practical use of this kind of background correction can only be recommended for FT.

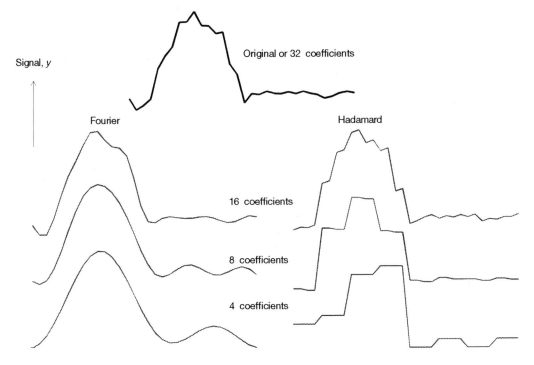

Fig. 3-12.
Transformation of signals from 32 data points and back-transformation by FT and HT using different numbers of coefficients. The numbers correspond to the remaining coefficients except the first coefficient, which represents the average (FT) or the sum (HT) of the spectrum

In Fig. 3-12 the results of forward and backward transformations of a spectrum are demonstrated. Even if only half of the original 32 coefficients are used, i.e. the 16 most important, the original data are quite well reproduced. When back-transformation is performed with only few coefficients the different base functions of the two transformations can be easily recognized, i.e. for FT the trigonometric and for HT the Walsh function.

In Fig. 3-12 a filtering effect can also be seen in connection with the reduction of coefficients. The difference from the above mentioned methods of signal filtering lies only in the concrete choice of the filter coefficients. The principle of back-transformation is the same for data filtering as that used here for data reduction.

Spline functions

In addition to the smoothing methods based on digital filters and on transformations, some further possibilities exist for signal smoothing. Among these are:

- *Local approximations:*
  Here the functional dependence or signal curve is split into intervals and these intervals are fitted piecewise, e.g., by straight-line models. Unfortunately, no smooth curves result but discontinuities emerge in the derivatives of the curve.
- *Modeling with known base functions:*
  If it is known, for example, that a spectroscopic band obeys the Lorentz function, then signal processing (smoothing, ana-lytical derivatives) is feasible by estimation of the para-meters of that base function. The parameters are commonly estimated by nonlinear regression analysis (cf. Sec 6.3.1).
  Typical problems with this method are that no unique base function can be found for the entire measurement range.
- *Spline functions:*
  These represent a compromise between a polygon trace and an interpolation polynomial of higher order. The main ad-vantage of spline functions is their differentiability in the entire measurement domain.

In the following text spline functions will be considered in more detail.

Interpolating splines

To construct an interpolating spline the x range is split into several intervals separated by so-called *knots*. The knots may be identical with the index points on the x variable axis.

An *interpolating* cubic spline $S(x)$ for observations on the abscissa grid $x_1 < x_2 < ... < x_n$ fulfils the following conditions:

Splines with variable knot loca-tions are termed *adaptive splines*.

- The spline $S(x)$ is interpolating, i.e. at the knots $k = 1,..., n$ the measured value, y_k, is equal to the spline value $S(x_k)$.
- Within the knots k, $S(x)$ obeys the continuity constraint on the function and on its twofold derivatives.
- $S(x)$ is a cubic function in each considered subrange $[x_k, x_{k+1}]$ for $k = 1,..., n - 1$.
- Outside the range from x_1 to x_n $S(x)$ is a straight line.

70

Fig. 3-13 shows a spline function that fulfils the above mentioned conditions.

The cubic function is defined by:

$$y = A_k (x - x_k)^3 + B_k (x - x_k)^2 + C_k (x - x_k) + D_k \quad (3\text{-}33)$$

where A_k, B_k, C_k and D_k are the spline coefficients at data point k.

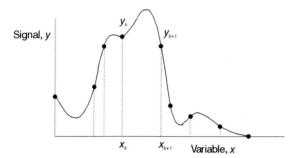

Fig. 3-13.
Interpolating spline

For a fixed interval between the data points x_k and x_{k+1} the following relationships are valid for the signal values and their derivatives:

$$y_k = D_k \qquad (3\text{-}34)$$

$$y_{k+1} = A_k (x - x_k)^3 + B_k (x - x_k)^2 + C_k (x - x_k)$$

$$y'_k = C_k$$

$$y'_{k+1} = 3A_k (x - x_k)^2 + 2B_k (x - x_k) + C_k$$

$$y''_k = 2B_k$$

$$y''_{k+1} = 6A_k (x - x_k) + 2B_k$$

By additional reshaping the spline coefficients can be determined from Eqs. 3-34.

Smoothing splines

The spline coefficients can also be determined by a method which also smoothes the data simultaneously. For this the ordinate values $\hat{y}_k$ are calculated such that the differences from the observed values y_k are positive proportional jumps r_k in their third derivative at point x_k:

$$r_k = f_k'''(x_k) - f_{k+1}'''(x_{k+1}) \qquad (3\text{-}35)$$

$$r_k = p_k (y_k - \hat{y}_k) \qquad (3\text{-}36)$$

The proportionality factors p_k are determined, e.g., by cross validation (cf. Sec. 5.2.1).

In contrast with polynomials, spline functions may approximate and smooth any kind of curve shape.

Problems, however, arise, if the intervals between the knots are not sufficiently narrow and the spline begins to oscillate (cf. Fig. 3-13). Also, in comparison with polynomial filters many more coefficients must be estimated and stored, since in each interval different coefficients apply. An additional disadvantage is valid for smoothing splines, where the parameter estimates are biased. The statistical properties of spline functions are, therefore, more difficult to describe than are those of linear regression (cf. Sec. 6.1).

3.2 Time – series analysis

Characterization of a set of measurements as a time series in the sense of a stochastic process is of interest in different ways, i.e.:

- Time-dependent monitoring of pH-values, or metal or ion concentrations in waters and soils
- Determination of the constituents of biological fluids over time, e.g. for monitoring blood glucose levels
- Description of the time-dependent stability of a spectroscopic source, e.g. the inductively coupled plasma in atomic emission spectroscopy
- Assessment of the performance of a continuously or discontinuously operating analyzer.

Relationship between variables can be described by means of *correlation* and *covariance*.

The methods for smoothing, derivation, integration or transformation as discussed in Sec. 3.1 can also be applied to a time series. In this section we learn about correlation methods. Correlations within a time series are described in terms of *autocorrelation* or *autocovariance*. Two different time series are characterized by means of *cross-correlation*.

Typical information to be derived from such models induces information about:

- Drift
- Noise
- Periodicity of processes, e.g. seasonal components
- Forecasting (prediction) of future values on the basis of the series history.

We start with correlations *within* a measurement series.

Autocorrelation and autocovariance

Correlations of data within a time series can be found if the data are plotted against successive values. Consider the time series given in Fig. 3-14, where the monthly recorded sulfur concentration per liter of snow, $y(t)$, is plotted against time, t.

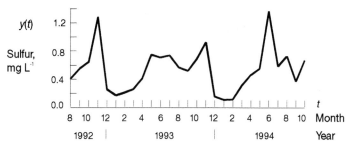

Fig. 3-14.
Time series for monthly recorded concentration of sulfur as sulfate

The correlations are obtained by plotting the measurement at time t, $y(t)$, against the value at time $t + 1$, i.e. $y(t + 1)$, at time $y(t + 2)$, or in general at time $y(t + \tau)$. τ represents the so-called *time lag*. In Fig. 3-15 the dependencies are plotted for the lags 0, 1, 7 and 12.

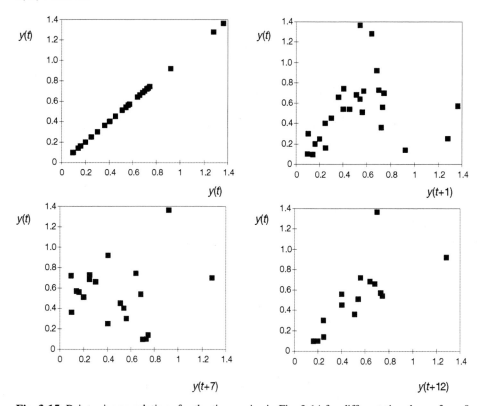

Fig. 3-15. Point-wise correlations for the time series in Fig. 3-14 for different time lags of $\tau = 0$, 1, 7 and 12 data points, with the correlation coefficients $r(\tau) = 1.000, 0.243, 0.0209$ and 0.365, respectively

A time lag of $\tau = 0$ represents the plot of the time series against itself. There correlation is complete. With increasing time lag, i.e., with increasing distance between the data points, the correlation is expected to decrease if no periodicities or drifts within the time series are present.

To measure the amount of correlation empirical autocorrelation is applied (cf. correlation coefficient according to Eq. 5-12 in Sec. 5.1). For autocorrelation of a function of n data points the *empirical autocorrelation, $r(\tau)$,* for time lag τ is defined by:

$$r(\tau) = \frac{\sum\limits_{t=1}^{n-\tau}(y_t - \bar{y})(y_{t+\tau} - \bar{y})}{\sum\limits_{t=1}^{n}(y_t - \bar{y})^2} \qquad (3\text{-}37)$$

Here $\bar{y}$ is the arithmetic mean. The expression in the denominator of Eq. 3-37 is a measure of the variance, s^2, because:

$$s^2 = \frac{\sum\limits_{t=1}^{n}(y_t - \bar{y})^2}{n-1-\tau} \quad \text{or} \quad \sum\limits_{t=1}^{n}(y_t - \bar{y})^2 = (n-1-\tau)s^2$$

A stochastic process is termed *stationary*, if the signal generating process is *time invariant*. All distributions and statistical parameters of a stationary process are independent of time. A process varying with time is called *non-stationary*.

Mean and variance are considered to be constant. This assumption is only true for a *stationary process*.

It should be mentioned, that the numerator in Eq. 3-37 contains the autocovariance (cf. Sec. 5.1, Eq. 5-10). The *empirical autocovariance* with time lag τ is defined by:

$$c(\tau) = \frac{1}{n}\sum\limits_{t=1}^{n-\tau}(y_t - \bar{y})(y_{t+\tau} - \bar{y}) \qquad (3\text{-}38)$$

If one calculates the autocorrelation coefficients for the correlations in Fig. 3-15, then when $\tau = 0$ we obtain for the correlation a value of 1, because:

$$r(0) = \frac{\sum\limits_{t=1}^{n}(y_t - \bar{y})(y_t - \bar{y})}{\sum\limits_{t=1}^{n}(y_t - \bar{y})^2} = 1 \qquad (3\text{-}39)$$

For increasing time lags from 1 to 7 the autocorrelations decrease to $r(1) = 0.243$ and $r(7) = 0.0209$, respectively. For $\tau = 12$ a higher value for the empirical autocorrelation again results, i.e. $r(12) = 0.365$. This fact points to periodicity in the time series.

In order to evaluate all possible autocorrelations, $r(\tau)$ is plotted against the time lag τ in a correlogram, also called an *autocorrelation function*. The autocorrelation function for the time series of sulfur concentrations in snow is given in Fig. 3-16. The time lags τ correspond here to monthly periods.

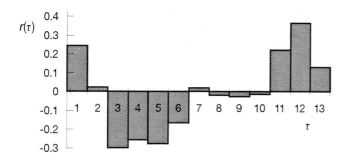

Fig. 3-16.
Autocorrelation function for the time series in Fig. 3-14. The correlation at $\tau = 0$ (cf. Fig. 3-15) is not drawn in the figure

At the beginning of the correlogram the correlations between the measurements decrease rapidly. For distances of 3 and more months the sulfur values result in negative correlations that might correspond to the four seasons (cf. Fig. 3-16).

Example 3-5: *Empirical autocorrelation*

For the time series of sulfur concentrations in snow (Fig. 3-14) the empirical autocorrelation must be calculated for the time lag $\tau = 12$. The 27 individual data are given in Table 3-3.

As the mean of the data a value of $\bar{y} = 0.53$ is obtained. By means of Eq. 3-37 the result for the autocorrelation is:

$$r(12) = \frac{\displaystyle\sum_{t=1}^{27-12}(y_t - \bar{y})(y_{t+12} - \bar{y})}{\displaystyle\sum_{t=1}^{27}(y_t - \bar{y})^2}$$

$$= \frac{\left[\begin{array}{c}(0.40-0.53)(0.56-0.53)+(0.54-0.53)(0.51-0.53)\\ +\ldots(0.684-0.53)(0.66-0.53)\end{array}\right]}{(0.40-0.53)^2 + (0.54-0.53)^2 + \ldots + (0.66-0.53)^2} = 0.3$$

The corresponding plot for the calculated value of 0.365 of the autocorrelation at $\tau = 12$ is given at the bottom right of Fig. 3-15.

Table 3-3.
Individual values for the time series in Fig. 3-14

t	Month/Year	$y(t)$
1	8/92	0.400
2	9/92	0.540
3	10/92	0.640
4	11/92	1.280
5	12/92	0.250
6	1/93	0.160
7	2/93	0.200
8	3/93	0.248
9	4/93	0.404
10	5/93	0.744
11	6/93	0.700
12	7/93	0.730
13	8/93	0.560
14	9/93	0.510
15	10/93	0.684
16	11/93	0.920
17	12/93	0.140
18	1/94	0.096
19	2/94	0.100
20	3/94	0.300
21	4/94	0.452
22	5/94	0.540
23	6/94	1.364
24	7/94	0.570
25	8/94	0.720
26	9/94	0.360
27	10/94	0.660

Autocorrelation functions for typical processes

For interpretation of the autocorrelation function it is useful to know the graphs of characteristic time series.

Uncorrelated data

The first step of data analysis should be a check to determine whether the data are uncorrelated or correlated. Uncorrelated data do not show any trends in their autocorrelation function (Fig. 3-17). Note in Fig. 3-17, how small the $r(\tau)$-values are for the empirical autocorrelations. Such data can be described by the methods discussed in Sec. 2. In other words, uncorrelated data are a prerequisite for application of the methods of descriptive statistics given in Chapter 2.

Fig. 3-17.
Autocorrelation function
of uncorrelated data

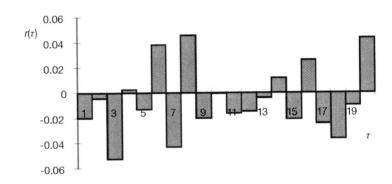

Correlated data

A random time series with low correlations between observations provides an autocorrelation function, as shown in Fig. 3-18. For a stationary process of the first order the function can be described by the following exponential model:

$$r(\tau) = e^{(-\tau/T)} \tag{3-40}$$

where T is the *time constant* of the process.

Fig. 3-18.
Autocorrelation function
for weakly correlated data
according to a first-order process.
The time series is based on glucose
determinations in urine over time

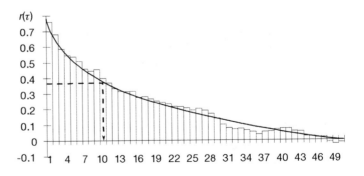

For determination of the time constant T nonlinear or, after logarithmic transformation, linear regression can be applied. The time constant can also be evaluated at the specific correlation coefficient $r(T) = 0.37$. At this point $\tau = T$, and therefore:

$$r(T) = e^{(-1)} = 0.37 \qquad (3\text{-}41)$$

For the data in Fig. 3-18 a time constant of $T = 10$ results from use of Eq. 3-41.

Random processes with drift and periodicities

Theoretically, fluctuation about zero is expected for a random process after a decrease of $r(t)$. For a *drifting* process the autocorrelation function remains at a positive or negative correlation level (cf. Fig. 3-19).

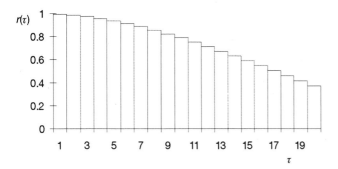

Fig. 3-19.
Autocorrelation function for a time series with drift, as found for measurements with a chemical sensor

The methods introduced for background correction can be used for correction of drifts (cf. Sec. 3.1).

Periodic processes show characteristic dependences as demonstrated in the example of the seasonal variations of sulfur concentrations in Fig. 3-16. Periodicities or fluctuations can be recognized and quantified from autocorrelation functions much better than from the time series.

Cross-correlation

To describe the correlation of two different time series, $y(t)$ and $x(t)$, the *empirical cross-correlation* for time lag τ is calculated by use of the following equation:

$$r_{xy}(\tau) = \frac{\sum\limits_{t=1}^{n-|\tau|} x_t y_{t+\tau}}{\sqrt{\sum\limits_{t=1}^{n} x_t^2 \sum\limits_{t=1}^{n} y_t^2}} \qquad (3\text{-}42)$$

Cross-correlation can be used to investigate the input-ouput behaviour of an analytical system or to compare a theoretical band shape with an observed one. The greatest correspondence between the two time series is observed in the correlogram at the position where a maximum for the empirical correlation is found.

Fig. 3-20.
Schematic demonstration of cross-correlation between an input signal x and and output signal y. The time lag can be derived from the shift in the correlation maximum.

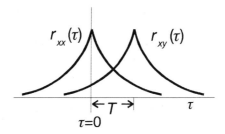

Autoregression

A close relationship exists between correlation of data and their regression on each other (cf. Sec. 6.1). It is, therefore, possible to model the successive data of a time series by a linear regression model in order to predict future values.

For *autoregression* of a time series we obtain:

$$y(t+\tau)-\bar{y} = r(\tau)[y(t)-\bar{y}]+e(t+\tau) \tag{3-43}$$

The measurement at time t plus the time lag τ is predictable on the basis of the autocorrelation coefficient, $r(\tau)$, and the y-value at time t. e represents the random error. Note that this is a very simple model. Good predictions can only be expected for time series that genuinely obey this simple autoregressive model.

Correlation methods are not restricted to the characterization of time-dependent measurements. They can also be used if correlations between spectra, chromatograms or other correlated analytical measurement series are to be investigated.

3.3 General reading

Boor, de C., *A Practical Guide to Splines*, Springer, 1978.
Bowerman, B. L., *Forecasting and Time Series*, Duxbury Press, Belmont, CA (USA), 1993.
Kalman, R. E., Journal of Basic Engineering **82** (1960) 35.
Rutan, S. C., *Fast On-line Digital Filtering*, Chemometrics Int. Lab. Sys. **6** (1989) 191.
Savitzky, A. and Golay, M. J. E., Anal. Chem. **36** (1964) 1627.
Vaseghi, S. V., *Advanced Signal Processing and Digital Noise Reduction*, Wiley-Teubner, Chichester, Leipzig, 1996.
Zupan, J., *Algorithms for Chemists*, Wiley, Chichester, 1989.

Questions and problems

1. Describe the useful and interfering frequencies in an analytical signal.
2. Explain the effect of filter-width on the noise and structure of a signal trace such as a spectrum or a chromatogram.
3. The Kalman filter is especially useful for real-time filtering. Why?
4. What are the benefits of signal derivation and which derivatives should by applied in which situation?
5. What are the base functions of FT and HT and for which applications is each particular transformation best used?
6. What is the difference between a polynomial and a spline function?
7. What information can be deduced from autocorrelation functions?

4 Optimization and experimental design

Learning objectives

- To provide an introductory course in systematic optimization methods in analytical chemistry
- To select the most important factors that influence a given analytical problem on the basis of the statistical approaches of experimental design, and evaluating the factor effects and their interactions by means of statistical tests
- To discuss the design of experiments for modeling the relationship between responses and factors and to apply response-surface methods for locating the optimum
- To search for the optimum by sequential methods, i.e. by means of the Simplex method of Nelder and Mead

Effective experimentation and the development of optimized methods are fundamental aims of any experimentor. Typical goals of the analyst are, in this connection:

- Investigation of factors which affect an analytical signal to judge the robustness of an analytical method or to test for interferences.
- Optimization of the performance of analytical methods with respect to quality criteria, such as precision, trueness, sensitivity, detection power or signal-to-noise ratio.
- Optimization of the composition of a digesting agent or of a chemically sensitive layer for the development of sensors.

Typical *factors* in analytical chemistry are the pH-value, reagent concentration, temperature, flow rate, solvent, eluent strength, mixture components, irradiation, atomization time or sputtering rate. Typical *responses* are the analytical figures-of-merit as well as objective functions that consist of combinations of different quality criteria. Objective functions and factors are considered in detail in Sec. 4.1.

In principal, two approaches in experimental optimization can be distinguished. First, selection and testing of the most important factors, and their subsequent optimization, are based on subjective experience of the experimentor or analytical expert. The success will be dictated then by the knowledge level of the domain expert. If the know-how is low then the workload might become large. In addition, as we will learn below, a nonsystematic approach does not guarantee discovery of the optimum.

A *systematic* optimization is always preferable to a trial-and-error approach.

Second, the investigation of the factors and their optimization can be performed systematically. This systematic approach is the subject of this chapter. The most successful experimentor will be the one who is able to support his expert-based policy with systematic investigations.

Systematic optimization

Systematic optimizations are carried out in the following sequence:
- Choice of an objective function
- Selection of the most important factors
- Optimization

Very often the optimization criterion is simply an analytical signal or the analysis time. In more complicated situations, however, objective functions that are composed of several criteria, such as selectivity, sensitivity, and precision have to be considered. Therefore, combination of objective criteria into a single function is an important topic in analytical chemistry.

Potential factors that affect a given objective function are best selected by the domain expert in the particular analytical field. The test for significance of the factors' influence has to be performed on the basis of a simple experimental design, a screening design, by means of statistical tests. Factors should not be kept or eliminated solely for subjective reasons.

To find the most suitable factor combinations we can distinguish between simultaneous and sequential optimization approaches.

Simultaneous methods

With *simultaneous* strategies the relationship between responses and factors is studied by running an experimental design, constructing a mathematical model, and investigating the relationship by use of so-called response surface methods (RSM). Very often RSMs are aimed at judging this relationship graphically and the consequences are drawn from those plots. If the optimal point is desired it can be found by calculating the partial derivatives with respect to the individual factors or by applying a grid search over the entire response surface.

Sequential methods

Simultaneous methods are based on a mathematical *model* of the response area, whereas sequential approaches represent *search methods*.

Sequential strategies of optimization are based on an initial experimental design followed by a sequence of further measurements in the direction of the steepest ascent or descent. That is, no quantitative relationship between factors and responses is evaluated but the response surface is searched along an optimal (invisible) path. The two strategies are exemplified in Fig. 4-1.

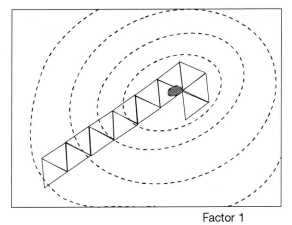

Factor 2

Factor 1

Fig. 4-1.
Response vs. factors plot.
With response surface methods
the response is described by a
mathematical model
(dotted contour lines of the
response surface). By use of search
methods the response is measured
along a search path, here along
a Simplex path (cf. Sec. 4.3).

4.1 Objective functions and factors

A prerequisite for any optimization is the definition of one
or several objective criteria (figures-of-merit). For computer-
aided or automatic optimizations the objective functions need
to be represented in a computer-readable format. In general the
following types of objective criteria can be distinguished:
- continuous or discrete continuous quantities, e.g. yield, time
  demand, analytical figures-of-merit or deviations between
  model and experiment;
- discrete (nominal scaled) quantities, e.g. the number of
  crystallizations or extractions;
- ordinal scaled values, e.g. sensory data, such as different
  degrees of sweetness of a raspberry jam, in the sense of a
  ranking order.

The sought optimum of an objective function is either the
minimum, e.g. minimum time demand, or the maximum, e.g.
the yield of a chemical reaction.

In analytical chemistry the analytical performance characteristics constitute ever important objective criteria that must
often be considered in combination.

Analytical performance characteristics

Calibration function

In connection with the performance of the calibration function several characteristics are combined, such as the sensitivity of an analytical method and its working range.

The *sensitivity* corresponds to the slope of the calibration curve. The dependence of the *calibration function* for the analytical signal, y, on the concentration (or mass), x, is:

$$y = b_0 + b_1 x \qquad (4\text{-}1)$$

The sensitivity, b_1, for this calibration function is then defined by (cf. Fig. 4-2):

$$b_1 = \frac{\Delta y}{\Delta x} \qquad (4\text{-}2)$$

The intercept b_0 represents an uncorrected blank or background value. If the measurement is obtained against a blank a model without the term b_0 would be valid.

Fig. 4-2.
Sensitivity for a straight line calibration of the functional dependence of signal on concentration

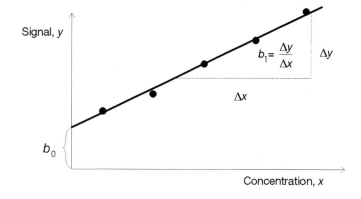

In case of *curved calibration gaphs* the sensitivity has to be reported together with the concentration considered.

In atomic absorption spectrometry the use of inverse sensitivity is commonly defined by the mass or concentration per 1% absorption (0.00436 absorbance units). The sensitivity can be given as a constant value for the whole concentration range only for a straight-line calibration curve. For curved calibration graphs the sensitivity varies with concentration and optimization is often performed with the aim of enhancing the linear range.

The *dynamic range* corresponds to the valid range of the functional dependence of the signal on concentration or mass. The *analytical* or *working range* denotes that interval between the lowest and highest concentration for which accurate measurements are feasible for evaluation of random and systematic errors. Outside this interval the measurements are considered uncertain.

Detection limit and limit of determination

The detection limit describes the minimum concentration that can be determined with a given analytical method. For evaluation of the *detection limit* the signal at the detection limit, y_{DL}, is calculated from the blank mean and the standard deviation as follows:

$$y_{DL} = y_B + 3s_B \qquad (4\text{-}3)$$

The factor of three provides sufficient statistical certainty to account for the errors which result from transformation from the signal to the concentration domain, for the necessary assumptions on the distribution of the blank values and for the limited number of measurements that only enable the determination of an estimate of the standard deviation of the blank value (s_B).

After reshaping of the calibration function in Eq. 4-1 the concentration at the detection limit, x_{DL}, is evaluated as:

$$x_{DL} = \frac{y_{DL} - b_0}{b_1} \qquad (4\text{-}4)$$

To characterize a boundary at which quantitative analysis is still feasible the limit of determination is used. The *limit of determination* is defined as the lowest analyte concentration that can be determined with acceptable accuracy.

Accuracy of analyses: precision and trueness

The accuracy of an analysis is determined by its precision and trueness. Precision corresponds to the fraction of random errors and trueness to that of systematic errors (bias). Fig. 4-3 illustrates the two types of error for measurements of a signal, y.

Accuracy is the above placed term over *precision* and *trueness*.

In the concentration domain we obtain:

$$e = \underbrace{(x - \bar{x})}_{\text{random error}} + \underbrace{(\bar{x} - x_{true})}_{\text{bias}} \qquad (4\text{-}5)$$

where e is the error determining the accuracy of the analysis, x is the determined concentration, $\bar{x}$ represents the concentration mean, and x_{true} is the true concentration.

Precision characterizes the repeatability of measurements. For n independent random measurements it can be described by the standard deviation as a measure of dispersion. In the concentration domain, x, is calculated by use of Eq. 2-4:

$$s = \sqrt{\frac{\sum_{i=1}^{n}(x_i - \bar{x})^2}{n-1}} \qquad (2\text{-}4)$$

Fig. 4-3.
Random and systematic errors for
measurements in the signal domain

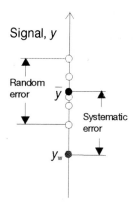

The *trueness* for characterization of systematic errors
is specified by the rate of recovery. This is the ratio of the
observed mean and the true value given as a percentage
by:

$$RR(\%) = \frac{\bar{x}}{x_w} 100 \qquad (4\text{-}6)$$

Specificity and Selectivity

The *selectivity* of an analytical method is a measure of
the extent to which the determination of an analyte suffers
from interference from accompanying analytes or matrix
components. A *fully selective* analytical method enables
the analytes to be selected for determination without interferences. One says the method is *specific* for the individual
analytes.

A prerequisite for undisturbed quantitative analysis is the
existence of interference-free signals for each component to be
determined. Signals related to several components are only
available for so-called two-dimensional analytical methods.
The signals are monitored for their dependence on a second
dimension, z, such as wavelength, time or electrode potential
(cf. Fig. 4-4). As a measure of the selectivity the analytical
resolution, N, is used. The *analytical resolution* is defined as
the ratio of the average signal of the two adjacent peaks on the
z-axis to the the signal at half-maximum, Δz:

$$N = \frac{z}{\Delta z} \qquad (4\text{-}7)$$

The larger the resolution the greater is the selectivity of the
method.

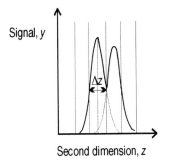 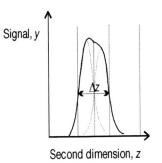

Signal, y   Signal, y

Δz Δz

Second dimension, z   Second dimension, z

Fig. 4-4.
Illustration of the analytical resolution for differently separated peaks or bands

If an analytical is completely selective, optimization of its selectivity would be superfluous. A baseline-separated chromatogram of all components of a sample is one example of a fully selective analytical method.

An unselective or partially selective method is characterized by overlapping analyte signals. The aim of optimization of such methods is improvement of their selectivity, if possible up to the development of a fully selective method. Often, however, signal overlap can only be reduced and the methods of multicomponent analysis have to be used to correct for the remaining interferences computationally (cf. Sec. 6.2.3).

To characterize incomplete selectivity, on the one hand, measures are applied that consider only two adjacent signals related to the corresponding two components. Table 4-1 provides selectivity measures from different areas of analytical chemistry. Multicomponent systems, e.g. chromatography, can also be described by appropriate combination of selectivity measures (cf. Table 4-2).

More general selectivity measures can be derived if the principles of multicomponent analysis are explored. Consequent application of the laws of error propagation enables selectivity criteria to be derived for the general case of multicomponent analysis that can even cope with different concentration ratios of analytes in the sample.

For characterisation of *limited selectivity* the notions interference, cross sensitivity or overlap are common.

Table 4-1.
Selectivity measures in analytical chemistry

Selectivity measure	Analytical principle	Formula
Selectivity coefficient, K_{ij}^{pot}	Potentiometry	$K_{ij}^{\text{pot}} = \dfrac{a_i}{a_j}$
Separation factor, α	Chromatography	$\alpha = \dfrac{k'_1}{k'_2}$
Resolution, R_s	Chromatography	$R_s = \dfrac{\sqrt{N}}{4}(\alpha-1)\dfrac{k'}{1+k'}$

a_i, a_j, activities of ions i or j; k'_i, capacity factor; N, plate number

Other figures-of-merit are the *signal-to-noise ratio* or the *signal-to-background* ratio. A measure of the signal-to-noise (S/N) ratio is the quotient of the signal means, $\bar{y}$, and the standard deviation of the signal, s_y:

$$\frac{S}{N} = \frac{\bar{y}}{s_y}$$

Since the principles of multicomponent analysis are not introduced until Sec. 6.2.3, the corresponding selectivity criteria will be discussed there.

Time, cost and risk

Minimization of the time demand or cost for analyses might also constitute objective functions. To optimize an entire analytical procedure methods of operational research might be needed in addition to the systematic approches considered in this section. This is especially important in cases where risk assessments are required in connection with the analytical procedure.

Accounting for several performance characteristics

In practice several objective criteria are very often important in optimization. The simultaneous optimization of an analytical method with regard to selectivity and time demands is a typical example for this situation.

Aggregation of performance characteristics

Multicriteria decision-making is feasible in two ways. First the different performance characteristics are aggregated to produce an objective function, most easily by use of a weighted sum. The objective function, Z, is obtained from the p individual objective criteria, z_i, by use of the equation:

$$Z = w_1 z_1 + w_2 z_2 + \ldots + w_p z_p = \sum_{i=1}^{p} w_i z_i \tag{4-8}$$

where w_i is the weight of objective criterion i.

The weights must be adjusted in such a way that they reflect the real influence of the performance characteristics on the total result. This is usually not easy to do. In addition, with this method the kind of aggregation must be decided in advance and optimization is carried out with regard to a single point.

Different aggregations of objective criteria have been developed for particular analytical methods. Table 4-2 gives examples of objective functions for chromatography and spectroscopy. The objective function for chromatograpy (CRF – chromatographic response function) takes into account all m peaks of the chromatogram, the time t for elution of the last peak, the noise, N_i, at the measurement point of peak i, and the selectivity of peak separation on the basis of Kaiser's measure of peak separation f/g (cf. Fig. 4-5). For optimal separations the CRF is maximized.

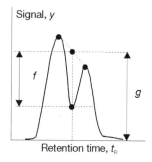

Signal, y

Retention time, t_R

Fig. 4-5.
Peak separation after Kaiser

For judging multi-element determination by means of atomic emission spectrometry (AES) or by X-ray fluorescence (XRF) analysis the evaluation of the signal-to-background ratio, I to I^B, for all p signals is proposed.

The advantages and disadvantages of the objective functions must be tested in the actual application of the particular method. A general disadvantage of the above mentioned methods is that the weighting of the objective criteria is done by means of a fixed weight.

Table 4-2.
Aggregation of performance characteristics to produce an objective function (explanation in text)

Objective function	Analytical method	Formula
CRF (chromatographic response function)	Chromatography	$CRF = \dfrac{1}{t} \prod\limits_{i=1}^{m-1} \dfrac{f_i}{g_i + 2N_i}$
Leary-criterion, Z_{AES}	Atomic emission spectrometry	$Z_{AES} = \dfrac{p}{\sum\limits_{i=1}^{p} \left(\dfrac{I_i}{I_i^B} \right)^{-1}}$
Signal-to-background ratio, Z_{XRF}	X-ray fluorescence analysis	$Z_{XRF} = \dfrac{1}{t} \sum\limits_{i=1}^{p} \dfrac{I_i - I_i^B}{I_i^B}$

A more flexible evalution is feasible by use of cost or utility functions. These can be constructed on the basis of conventional mathematical functions or by means of fuzzy functions (Sec. 8.3).

The preliminary decision on the aggregation of the criteria does not enable the discovery to find compromise solutions. The latter is possible if methods of polyoptimization are applied.

Polyoptimization

To find compromise solutions the entire region is investigated without aggregating the individual criteria in advance.

A prerequisite here is the feasibility of describing the relationship between the objective criteria and the factors by use of a mathematical model. The objective criteria can then be computed for all the factor combinations, and if not more than three criteria are to be considered they can be plotted or computationally investigated. In that way the compromise set of all Pareto optimal points can be found (cf. Fig. 4-6). Pareto optimal points are those points of factor combinations for which a change of one of the objective criteria would result in worsening of at least one other criterion.

Fig. 4-6.

Goal region for the simultaneous investigation of two objective criteria, e.g. selectivity and analysis time. The bold line characterizes Pareto optimal points

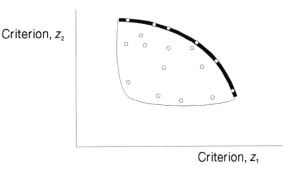

Criterion, z_2

Criterion, z_1

As can be seen in Fig. 4-6, the objective criteria cannot assume any arbitrary value. Depending on the experimental constraints on the factors the objective criteria lie in a bounded region. On the basis of the mathematical model the objective region can be investigated by calculation of the criteria at as many points as needed.

4.2 Experimental design and response surface methods

4.2.1 Fundamentals

Ceteris-paribus-principle

Optimization of an analytical problem or of an analytical procedure must be carried out by studying a limited number of factors. Very often it is easy to select the most important factors from deep knowledge about a given problem, e.g., for developing a new spectrophotometric method the factors pH value and reagent concentration would have to be studied, or in HPLC important factors are the constituents of the mobile phase and their concentrations.

Sometimes the effect of a factor can only be presumed and its effect would have to be assured by suitable screening experiments. Because of the complexity of most analytical problems there will be additional factors that are either unknown or that cannot be controlled by the experimenter. Uncontrolled factors might be the impurity of reagents, contamination of an electrode surface, the instability of a plasma source, or the changing quality of a laboratory assistant's work.

In a *screening experiment* the factor range is evaluated for the subsequent studies.

Since the study of all potential factors is usually prohibitive the effects of selected factors are investigated and the remaining factors are kept as constant as possible. This general principle is known as the *ceteris-paribus-principle*.

Replication

Replication of measurements for a given combination of factors is necessary for estimation of the experimental error. Furthermore, the error can be reduced by taking replicate measurements and averaging, i.e., by a factor of $1/\sqrt{n}$ if n measurements have been obtained.

Randomization

Randomization means running the experiments in a random order. Randomized experiments are obligatory if systematic errors (bias) cannot be avoided and must be detected. Imagine the construction of a calibration graph. If the concentrations are measured in ascending order it would not be possible to detect a systematic error caused by a positive drift of the signal with time. All experimental observations might lie on a straight line but the measured slope of the calibration graph is shifted systematically from the true slope (Fig. 4-7A). On the other hand, if the concentrations are measured in a random sequence, then you will notice that something is wrong because the variation of the observations will be exceptionally high (Fig. 4-7B).

In this example we have studied only one factor – the component concentration. When studying several factors randomization is likewise necessary. In multi-factor experiments the experimental design will have to be run in a randomized order as we will see below.

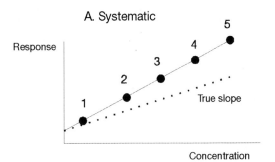

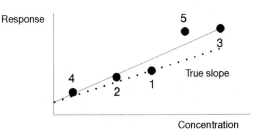

Fig. 4-7.
Experimental observations for construction of a calibration graph in systematic (A) and randomized order (B)

One of the basic prerequisites of all statistical tests is the existence of random and independent data. This assumption will also be valid for statistical tests in connection with experimental designs. Therefore, randomized experimentation is obligatory in this context. Carrying out the experiments in a random sequence will be one of the suppositions for being able to measure independent, uncorrelated and usually normally distributed data.

Random sequences should be read from random number tables or nowadays from a random number generator. This guarantees genuine randomness of the sequence instead of subjective selections.

Blocking

Uncontrolled factors lead to higher experimental errors. This increased error will also reduce the sensitivity of the experiments with respect to the factors to be studied. Therefore, the experiments should be designed such that uncontrolled factors can be detected and, by means of a further step, can be kept constant or eliminated.

There are two categories of uncontrolled factors. Uncontrolled influences might arise from either *unknown factors* or from *known factors that cannot be controlled*. The eventual impurity of a reagent is an example of an unknown factor. The changing quality of a laboratory assistant's work is a factor that is difficult to control. No account can be taken of completely unknown factors.

Known or presumably known factors can be detected by blocking the experiments. The idea is to run the experiments in blocks that show a minimum experimental variance within one block. For example, if a systematic investigation requires 12 experiments and you can run only four experiments a day the experiments should be arranged in three blocks with four experiments each day. Day-to-day effects could then be detected by considering the block effects with an adequate mathematical model as given below.

Of course, the experiments within a block should be run at random. Randomized experimentation with regard to some factors, and blocking of the experiments with regard to some other factors exclude each other. Therefore, in practice a compromise between randomized runs and blocked experiments must be found.

Factorial experiments

Factorial experiments are based on varying all factors simultaneously at a limited number of factor levels. This kind of experimentation is especially important in the beginning of an experimental study, where the most influential factors, their ranges of influence and factor interactions are not yet known. Factorial experiments enable experiments to be performed in the whole range of the factors' space. They reveal high precision with minimum experimental effort and they enable detection of factor interactions such as the dependence of enzyme activity on both pH-value and co-enzyme concentration.

Confounding

Confounding of parameter estimations for different factors occurs if the factor combinations are highly correlated and, therefore, no difference between the factor effects can be detected. Confounding is highly dependent on the specific experimental design. If, for example, the levels of two factors are changed in a constant ratio it would not be possible to distinguish between the effects of those two factors.

Symmetry

Factorial experiments should be partitioned in a balanced manner in the whole factor space. The same is true for replications in the experimental space. One reason for performing symmetric experiments is the avoidance of confounded factor effects. In addition, symmetric experiments might simplify data evaluation.

4.2.2 Two-level designs: screening designs

Full factorial designs

Designs on the basis of two levels for each factor are called screening designs. The most general design is a *full factorial design* at two levels. These designs are described as 2^k-designs where the base 2 stands for the number of factor levels and k expresses the number of factors.

Example 4-1: *2^3-Design*

As an example, a two-level three factorial design, 2^3, is given in Table 4-3 and Figure 4-8. The factor levels are scaled here to −1 for the lower level and +1 for the higher level. Other coding schemes are also used, e.g., 0 and 1, or − and +, or A and B for the low and high levels, respectively. One advantage of working with scaled factor levels is the feasibility of applying the same design to several investigations. Another advantage can be seen if we model the relationship between responses and factors quantitatively. With coded levels the size of the parameters will become comparable. Of course, the original variables can also be used in a study.

93

Table 4-3.

Full factorial design at two levels, 2^3 design. The star labeled experiments represent the half-fraction factorial design of Fig. 4-8.

Experiment	Factors			Response
	x_1	x_2	x_3	
1	−1	−1	−1	y_1
2*	+1	−1	−1	y_2
3	+1	+1	−1	y_3
4*	−1	+1	−1	y_4
5*	−1	−1	+1	y_5
6	+1	−1	+1	y_6
7*	+1	+1	+1	y_7
8	−1	+1	+1	y_8

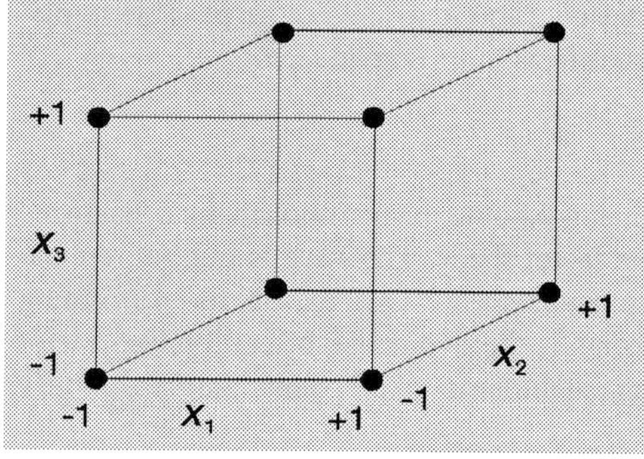

Fig. 4-8.

Full factorial design at two levels, 2^3 design. x_1, x_2 and x_3 represent the three factors

Fractional factorial designs

As long as the number of factors is small, full factorial designs can easily be run. For high factor numbers, however, the number of experiments will increase dramatically. For example, for the study of seven factors in a 2^7-design 128 experiments in total will be necessary. At this point we have to discuss the objectives of running a factorial experimental design. One reason is the estimation of factor effects. As we will see in the section on linear models (Sec. 6.2) it will not be necessary to evaluate the effects of seven factors by running 128 experiments. Another aim is to model the dependence of the responses on the factors. For modeling dependences of responses on two factor levels we would use a first-order polynomial. So in our case we would have to estimate seven parameters linked to the effects of the seven factors and perhaps an additional parameter that models the intercept on the ordinate axis. For computing this statistical model we deduce $128 - 8 = 120$ degrees of freedom which are obviously too many to test for the adequacy of a simple first-order polynomial model.

The number of experiments can be reduced if *fractional factorial designs* are used. For fractional factorial designs the number of experiments is reduced by a number p according to a 2^{k-p} design. For $p = 1$, so-called half-fraction designs result.

Example 4-2: 2^{3-1}-*Design*

For the example of three factors at two levels (cf. Fig. 4-8) the half-fraction design consists of $2^{3-1} = 4$ experiments as given in Fig. 4-9. In Table 4-4 the points of this design are labeled by an asterisk. The experiments can be run either at the given points or at the complementary corner points.

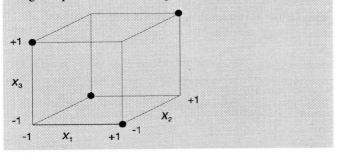

Fig. 4-9.
Fractional factorial design at two levels, 2^{3-1} half-fraction design, x_1, x_2 and x_3 represent the three factors

Table 4-4.
Factorial design for four factors at two levels. 2^{4-1} – half-cell design.

Run	Factor			
	x_1	x_2	x_3	x_3
1	−1	−1	−1	−1
2*	+1	−1	−1	+1
3	−1	+1	−1	+1
4*	+1	+1	−1	−1
5*	−1	−1	+1	+1
6	+1	−1	+1	−1
7*	−1	+1	+1	−1
8	+1	+1	+1	+1

Table 4-5.
Factorial design for five factors at two levels. 2^{5-1} – half-cell design

Run	Factor				
	x_1	x_2	x_3	x_4	x_5
1	−1	−1	−1	−1	+1
2	+1	−1	−1	−1	−1
3	−1	+1	−1	−1	−1
4	+1	+1	−1	−1	+1
5	−1	−1	+1	−1	−1
6	+1	−1	+1	−1	+1
7	−1	+1	+1	−1	+1
8	+1	+1	+1	−1	−1
9	−1	−1	−1	+1	−1
10	+1	−1	−1	+1	+1
11	−1	+1	−1	+1	+1
12	+1	+1	−1	+1	−1
13	−1	−1	+1	+1	+1
14	+1	−1	+1	+1	−1
15	−1	+1	+1	+1	−1
16	+1	+1	+1	+1	+1

95

Apart from saturated designs, numerous other fractional factorial designs can be used if the effect of many factors is to be studied. Specific designs can be taken from tables (cf. Table IX in the appendix) and can be generated with most of the software packages given in the appendix. Special designs for estimating only the main effects have been tabulated by Plackett and Burman. As an example their fractional factorial design for 11 factors at two levels is represented in Table 4-6.

Table 4-6.
Plackett and Burman fractional factorial design for estimating the main effects of 11 factors at two levels

Run	Factors											Response
	x_1	x_2	x_3	x_4	x_5	x_6	x_7	x_8	x_9	x_{10}	x_{11}	
1	+	+	−	+	+	+	−	−	−	+	−	y_1
2	−	+	+	−	+	+	+	-	−	−	+	y_2
3	+	−	+	+	−	+	+	+	−	−	−	y_3
4	−	+	−	+	+	−	+	+	+	−	−	y_4
5	−	−	+	−	+	+	−	+	+	+	−	y_5
6	−	−	−	+	−	+	+	−	+	+	+	y_6
7	+	−	−	−	+	−	+	+	−	+	+	y_7
8	+	+	−	−	−	+	−	+	+	−	+	y_8
9	+	+	+	−	−	−	+	−	+	+	−	y_9
10	−	+	+	+	−	−	−	+	−	+	+	y_{10}
11	+	−	+	+	+	−	−	−	+	−	+	y_{11}
12	−	−	−	−	−	−	−	−	−	−	−	y_{12}

A special case of a two-level factorial design is the *Latin square design*, which was introduced very early on to eliminate more than one blocking variable. A Latin square design for two factors is given in Table 4-7 along with the representation as a fractional factorial design.

Table 4-7.
2×2 Latin square design in different representations

Conventional representation:
4×4 Latin square

Blocking variable 1	Blocking variable 2	
	low	high
low	A	B
high	B	A

Factorial representation:

Blocking variable 1	Blocking variable 2	Blocking variable 3
low	low	A
low	high	B
high	low	B
high	high	A

Example 4-3: *Latin square*

Four formulations of a drug are to be studied with regard to bioequivalence by treating four subjects for four periods. For this a 4 × 4 latin square is constructed as given in Table 4-8. The formulations are coded by A, B, C and D.

Table 4-8.
4 × 4 Latin square

Subject No.	Period 1	2	3	4
1	A	B	D	C
2	D	C	A	B
3	B	D	C	A
4	C	A	B	D

Note the good balance of the latin square. All four drugs are given at each period to each person.

Estimation of factor effects

Screening designs are mainly used to estimate the effects of factors in an analytical investigation on a statistical basis. To understand the test procedure we will consider an example from kinetic-enzymatic determinations.

Example 4-4: *Factor effects estimation*

The determination of the enzyme ceruloplasmin on the basis of spectrophotometric measurements of the initial rate of *p*-phenylenediamine oxidation is investigated at a constant enzyme concentration of 13.6 mg L^{-1}. In the first step, a screening 2^3-design is used to study the effects of the factors pH-value, temperature, and the concentration of *p*-phenylenediamine.

The factor levels are given in Table 4-9 along with the experimental design and the measured initial rates. On the basis of the 2^3-design both *main* factor effects as well as *interactions* can be studied. The levels for factor interactions are calculated as products of the actual factor level combinations.

Table 4-9.
2^3- screening design and factor levels for estimation of the factors pH-value, temperature (*T*), and *p*-phenylendiamine (PPD) concentration

Factor	Level −1	+1
T, °C	35	40
PPD, mM	0.5	27.3
pH	4.8	6.4

Table 4-9 (continued)

Run	Coded factor levels						y, min^{-1}
	main effects			interaction effects			
	T	PPD	pH	T × PPD	T × pH	PPD × pH	
1	+1	−1	−1	−1	−1	+1	6.69
2	+1	+1	−1	+1	+1	−1	11.71
3	+1	+1	+1	+1	−1	+1	14.79
4	+1	−1	+1	−1	+1	−1	8.05
5	−1	−1	−1	+1	−1	+1	6.33
6	−1	+1	−1	−1	+1	−1	11.11
7	−1	+1	+1	−1	−1	+1	14.08
8	−1	−1	+1	+1	+1	−1	7.59

The factor effects are calculated as the absolute difference, |D|, between the responses of a factor at high and low levels. For example, for the factor, x_1, of the full factorial 2^3- design in Table 4-4 we obtain

$$D_{x_1} = \frac{y_2 + y_4 + y_6 + y_8}{4} - \frac{y_1 + y_3 + y_5 + y_7}{4} \qquad (4\text{-}9)$$

These differences are then tested against the experimental error expressed by the standard deviation, s, multiplied by the Student's t-value:

$$|D| \geq t(P, f)s \qquad (4\text{-}10)$$

$$
\begin{aligned}
D_T &= \frac{y_1 + y_2 + y_3 + y_4}{4} - \frac{y_5 + y_6 + y_7 + y_8}{4} \\
&= \frac{6.69 + 11.71 + 14.79 + 8.05}{4} \qquad (4\text{-}11) \\
&\quad - \frac{6.33 + 11.11 + 14.08 + 7.59}{4} = 0.53
\end{aligned}
$$

$$
\begin{aligned}
D_{\text{PPD}} &= \frac{11.71 + 14.79 + 11.11 + 14.08}{4} \\
&\quad - \frac{6.69 + 6.33 + 8.05 + 7.59}{4} = 5.76
\end{aligned}
$$

$$
\begin{aligned}
D_{\text{pH}} &= \frac{14.79 + 14.08 + 8.05 + 7.59}{4} \\
&\quad - \frac{6.69 + 6.33 + 11.71 + 11.11}{4} = 2.17
\end{aligned}
$$

$$
\begin{aligned}
D_{\text{T×PPD}} &= \frac{11.71 + 14.79 + 6.33 + 7.59}{4} \\
&\quad - \frac{6.69 + 8.05 + 11.11 + 14.08}{4} = 0.123
\end{aligned}
$$

$$D_{\text{T}\times\text{pH}} = \frac{14.79 + 8.05 + 6.33 + 11.11}{4}$$
$$- \frac{6.69 + 11.71 + 14.08 + 7.59}{4} = 0.053$$

$$D_{\text{PPD}\times\text{pH}} = \frac{6.69 + 14.79 + 6.33 + 14.08}{4}$$
$$- \frac{11.71 + 8.05 + 11.11 + 7.59}{4} = 0.858$$

With a standard deviation of 0.24 and a degree of freedom of $f = 3$ measured at the factor levels of run 3 we calculate for the experimental error:

$$t(0.95,3)\, s = 3.18 \times 0.24 = 0.76 \tag{4-12}$$

Comparison of the experimental error with the absolute differences reveals that the main factors pH and PPD concentration show a significant effect (D_{PPD} and D_{pH} are higher than 0.76) while the effect of the temperature can be neglected in the range studied (between 35 and 40 °C ($D_{\text{T}} < 0.76$)).

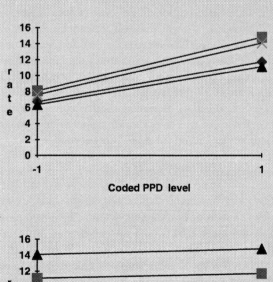

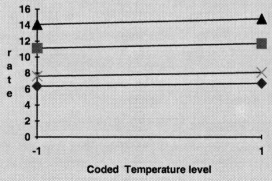

Fig. 4-10.
Factor effects in the kinetic-enzymatic oxidation of p-phenylenediamine by the enzyme ceruloplasmin

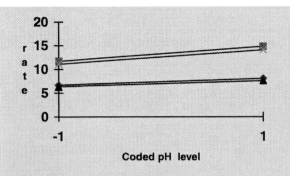

Coded pH level

Graphical inspection of the main factor effects can be carried out by study of Fig. 4-10. As found by calculation, there is minimal influence by the temperature. If the enzymatic reaction is run at the normal temperature of 37 °C the method will be rugged against temperature fluctuations. Notice that all measured initial rates are slightly higher at 40 °C than at 35 °C. Compared with the general experimental error, however, this effect has been found to be statistically insignificant.

The enhancing effect of the substrate PPD concentration and of the pH value on the rate can also be seen in Fig. 4-10. As a general rule the main effects will give parallel straight lines in the factor-effects plot. In the case of factor interactions the slope of the straight line will differ if the levels of the alien factors change. The latter is the case for interactions of the pH and PPD factors. As can be seen from Fig. 4-10 the changes of rates between the lower and higher levels of PPD concentration are more pronounced, if the factor pH is at the high level (points ■ and x). A similar effect is observed if the dependence of rate changes on pH are compared at high (points ■ and x) and low (points ◆ and ▲) PPD concentrations.

The calculated interaction effects are significant for the interaction of substrate PPD and pH value ($D_{PPD \times pH} > 0.75$).

In contrast to the simultaneous factorial design study, experimentation by variation of one variable at a time is limited to the estimation of the main effects, and no interactions, which are common in analytical chemistry, can be found. What cannot be evaluated with screening designs are curved dependences, i.e., designs at three or more factor levels are needed for more complicated relationships between responses and factors.

4.2.3 Three-level designs: response surface designs

In order to describe the relationship be, een responses and factors quantitatively we will use below mechanistic (physicochemical) or empirical models, e.g., polynomial models. These mathematical models should be equally able to describe both linear and curved response surfaces. Curved dependences can be modelled if the factor levels have been investigated at at least three levels.

Three-level factorial designs are known, therefore, as response surface designs. Full factorial three-level designs can be formalized in the same way as for two-level designs, i.e., a 3^k-design means k factors at 3 factor levels. Fig. 4-11 gives an example of a 3^2-design.

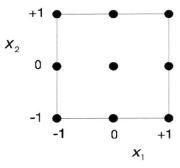

Fig. 4-11.
Full factorial three-level design for two factors, 3^2- design

Full factorial three-level designs are sometimes used for investigating few factors (two or three) although their statistical properties with regard to symmetry or confounding of parameter estimates are less favourable than those known for the two-level designs. With many factors the same problem arises as with two-level designs, i.e., the number of experiments becomes very high. These disadvantages led to the development of so-called optimal designs of which the *central composite design* and the *Box-Behnken design* are the most important.

The *number of degrees of freedom* of a factorial design is the number of runs minus the number of independent factor combinations.

Central composite design

Composite designs consist of a combination of a full or fractional factorial design and an additional design, often a star design. If the centers of both designs coincide they are called central composite designs. Consider a design that consists of a full factorial two-level design linked to a star design. For the number of runs, r, we obtain:

$$r = 2^{k-p} + 2k + n_0 \qquad (4\text{-}13)$$

where

k – number of factors
p – number for reduction of the full design
n_0 – number of experiments in the center of the design

Example 4-5: *Runs in a central composite design*

For a design with three factors one obtains for the number of experiments, r, according to Eq. 4-13:

$$r = 2^3 + 2*3 + 1 = 15$$

Fig. 4-12.

Central composite design
for three factors

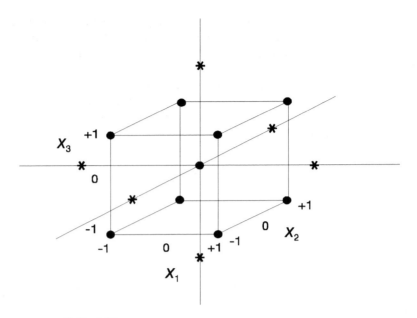

Table 4-10.

Central composite design for three factors consisting of a full factorial
two-level design and a star design

Run	Factors			Response
	x_1	x_2	x_3	
2^3-kernel design				
1	−1	−1	−1	y_1
2	+1	−1	−1	y_2
3	+1	+1	−1	y_3
4	−1	+1	−1	y_4
5	−1	−1	+1	y_5
6	+1	−1	+1	y_6
7	+1	+1	+1	y_7
8	−1	+1	+1	y_8
$2k$- star points				
9	−α	0	0	y_9
10	+α	0	0	y_{10}
11	0	−α	0	y_{11}
12	0	+α	0	y_{12}
13	0	0	−α	y_{13}
14	0	0	+α	y_{14}
center point				
15,16,17	0	0	0	y_{15}, y_{16}, y_{17}

A complete three-factor central composite design is depicted
in Table 4-10 and Fig. 4-12. The distance of the star points α
from the center can be different. For a uniformly rotatable
design

$$\alpha = 2^{(k-p)/4} \tag{4-14}$$

e.g., for 3 factors $\alpha = 2^{(3-0)/4} = 1.682$.

To estimate the experimental error, replications of factor combinations are necessary. Usually the center point is run thrice. The total number of runs in a central composite design with three factors then amounts to 17.

Replication here means carrying out a given factor combination several times.

Apart from the good statistical properties of the central composite design there is one experimental disadvantage. Because the star points are outside the hypercube the number of levels that must be adjusted for every factor is actually five instead of three in a conventional three level design. If the adjustment of levels is difficult to achieve an alternative response surface design would be the design introduced by Box and Behnken.

Box-Behnken design

In a Box-Behnken design the experimental points lie on a hypersphere equidistant from the center point as exemplified for a three-factor design in Fig 4-13 and Table 4-11. In contrast to the central composite design the factor levels have only to be

Table 4-11.
Box-Behnken design for three factors

Run	Factors			Response
	x_1	x_2	x_3	
1	+1	+1	0	y_1
2	+1	−1	0	y_2
3	−1	+1	0	y_3
4	−1	−1	0	y_4
5	+1	0	+1	y_5
7	−1	0	+1	y_7
8	−1	0	−1	y_8
9	0	+1	+1	y_9
10	0	+1	−1	y_{10}
11	0	−1	+1	y_{11}
12	0	−1	−1	y_{12}
13,14,15	0	0	0	y_{13}, y_{14}, y_{15}

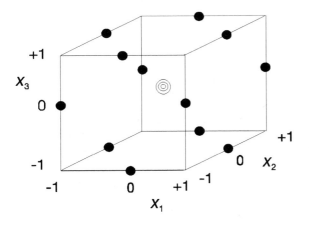

Fig. 4-13.
Box-Behnken design for three factors

adjusted at three levels. In addition, if two replicates are again performed in the center of the three-factor design the total number of experiments is 15 compared with 17 with the central composite design.

In a Box-Behnken design the experimental points lie on a sphere rather than on a cube.

In general, the Box-Behnken design requires few factor combinations, as is given in Table 4-12. The use of the design will be further explained in the chapter on response surface methods (below).

Table 4-12.
Number of factors and experimental points for the Box-Behnken design with three replicates in the center of each design

Number of factors	Number of experiments
3	15
4	27
5	46
6	54
7	62

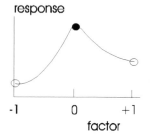

response

-1   0   +1

factor

One disadvantage of Box-Behnken designs might be the representation of the dependence of the responses in dependence on a single factor. This is because the corner points of the cube have not been measured but have to be computed by an appropriate response surface model.

Mixture designs

A special problem arises if there are additional relationships between the factors in an analytical investigation. This is true if mixtures, such as eluents in liquid chromatography or formulations in the pharmaceutical and textile industries, are under investigation. The constituents of a mixture given in portions by weight, volume or mole are confined to the assumption that the amounts of the N constituents sum up to 100% or (normalized) to 1, i.e.,

$$\sum_{i=1}^{N} x_i = 1 \quad \text{for} \quad x_i \geq 0 \tag{4-15}$$

The most important mixture design for an analyst is based on a so-called (k,d)-lattice. For the k factors the lattice describes all experimental points having the factor levels 0, $1/d$, $2/d$,...,$(d-1)/d$ or 1. In total the (k,d)-lattice has the following number of points:

$$\binom{d+k-1}{d} \tag{4-16}$$

Example 4-6: Lattice design

Consider a (3,3)-lattice design as given in Fig. 4-14. The number of points is calculated from

$$\binom{3+3-1}{3}=\binom{5}{3}=\frac{5!}{(5-3)!3!}=\frac{5\cdot4\cdot3\cdot2\cdot1}{2\cdot1\cdot3\cdot2\cdot1}=10$$

The factor levels are 0, 1/3, 2/3, 1, i.e. 0, 0.33, 0.66, 1.00, respectively. In HPLC the three constituents could be the solvents methanol, acetonitrile and water in the mobile phase.

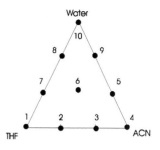

Ternary mixtures of water, THF (tetrahydrofuran) and ACN (acetonitrile) as mobile phase compositions in HPLC.

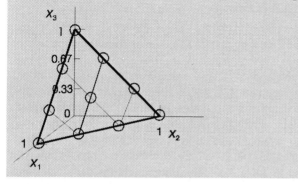

Fig. 4-14.
Mixture design for three factors at 10 levels based on a (3,3)-lattice design

RSM – response surface methods

Response surface methods are very useful for quantifying and interpreting the relationships between responses and factor effects. In analytical chemistry the relationships can be based on physical or physico-chemical models that are generalized by statisticians as so-called *mechanistic* models. Another way is empirical modeling where the parameters have no mechanistic meaning. General empirical models are second-order polynomials, where the response y is related to the variables (factors) x as follows:

$$y=b_0+\sum_{i=1}^{k}b_i x_i+\sum_{1\le i\le j}^{k}b_{ij}x_i x_j+\sum_{i=1}^{k}b_{ii}x_i^2 \qquad (4\text{-}17)$$

with

k –   number of variables (factors)
b_0 –   intercept parameter
b_i, b_{ij}, b_{ii} –   regression parameters for linear, interaction and quadratic factor effects

To estimate all parameters in Eq. 4-17 the experiments must be carried out at three factor levels, as discussed above for the response surface designs. The responses of experiments based on two factor levels can also, of course, be explored by RSM. However, no parameter estimates can then be obtained for curved (quadratic) factor effects; only linear and interaction effects can be visualized.

105

The estimation of the empirical parameters is a general problem of least-squares estimation by linear models. The basics are introduced in Sec. 6.2. With the parameters in hand the model can be used to plot the dependence of the response on the individual factors. The response surface method is explained here by further exploring the enzymatic determination of ceruloplasmin from the previous section on screening designs.

Small deviations from the factor levels −1, 0 and +1 will not significantly alter the statistical properties of a design.

Example 4-7: *Response surface methods*

In the above study the factors pH value and PPD were found to have a significant influence on the enzyme-catalyzed oxidation of the substrate PPD. To study the relationship between the response (initial reaction *rate*) and the significant factors quantitatively a design at three levels, a Box-Behnken design, is run at a temperature of 37°C. The concentration of the enzyme ceruloplasmin (CP) is included as a third factor in order to investigate the response characteristics of the analyte ceruloplasmin. Details of the Box-Behnken experiments are outlined in Table 4-13. The three levels for each factor are given along with the specific experimental design and the results of rate measurements.

Table 4-13.
Factor levels and Box-Behnken design for studying the determination of ceruloplasmin by RSM

Factor	Level		
	−1	0	+1
PPD, mM	0.5	14.3	27.3
pH	4.8	5.6	6.4
CP, mg L^{-1}	0.7	13.6	26.0

Run		Factors			Rate y,
Systematic	Randomized	PPD	pH	CP	min^{-1}
1	2	+1	+1	0.02	20.67
2	12	+1	−1	0.02	12.67
3	13	−1	+1	0.02	8.21
4	4	−1	−1	0.02	6.58
5	6	+1	0	+1	37.2
6	7	+1	0	−1	5.27
7	1	−1	0	+1	14.95
8	9	−1	0	−1	1.87
9	11	0.03	+1	+1	33.63
10	14	0.03	+1	−1	4.4
11	10	0.03	−1	+1	26.02
12	15	0.03	−1	−1	1.06
13	5	0.03	0	0.02	23.86
14	3	0.03	0	0.02	24.43
15	8	0.03	0	0.02	23.29

Parameter estimation with a second- order polynomial reveals the following final model:

$$y = 2.13 + 5.48 \text{ PPD} + 2.54 \text{ pH} + 1.23 \text{ CP}$$
$$+ 4.73 \text{PPD} \times \text{CP} - 6.32 \text{ PPD}^2 - 5.02 \text{ pH}^2 \qquad (4\text{-}18)$$

From the empirical model, it can be concluded that the three factors pH, PPD, and the enzyme concentration (CP) show main and quadratic effects (cf. Eq. 4-18) on the rate of the reaction. In addition, there is a statistically significant interaction between the factors substrate (PPD) and enzyme (CP). In analytical terms this means that for determinations of the enzyme the substrate concentration should be kept constant with high precision.

On the basis of the mathematical model the response surfaces can be explored graphically. An example plot of the dependence of the response rate on PPD concentration and pH is seen in Fig. 4-15A. The curved dependences in the direction of both factors lead to a maximum rate at coded levels of PPD of about 0.4 and at a pH of 0.2. This relates to decoded levels of 16.6 mM PPD and a pH value of 5.95. Maxima are best found from the contour plots as represented in Fig. 4-15.

A

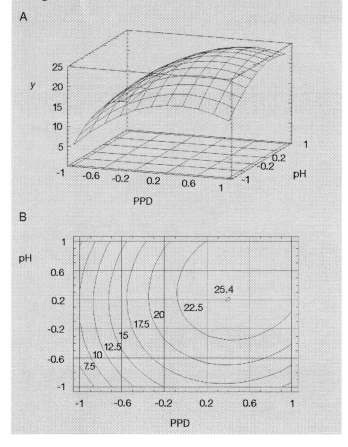

B

Fig. 4-15.
Surface (A) and contour (B) plots of rate of enzymatic degradation versus the factors PPD concentration and pH. The contour lines in B represent iso-rate lines

Coding of factor levels

To convert factor levels between original and coded levels the corresponding formulas will be given here. The coded level in the interval −1 to +1, x_i^*, is calculated from the original factor level, x_i, from:

$$x_i^* = \frac{x_i - M}{H} \qquad (4\text{-}19)$$

Here is M the mid-range and H the half-range of factor i:

$$M = \frac{high + low}{2} \qquad (4\text{-}20)$$

$$H = \frac{high - low}{2} \qquad (4\text{-}21)$$

Decoding of a factor level is carried out by converting Eq. 4-19 to the original factor level:

$$x_i = x_i^* \cdot H + M \qquad (4\text{-}22)$$

Example 4-8: *Coding and decoding of factors*

A. For PPD concentrations in the range from 0.5 to 27.3 mM the coded level for a concentration of 14.3 mM is to be determined (cf. Table 4-13). The mid-range and half-range are obtained according to Eqs. 4-20 and 4-21:

$$M = \frac{27.3 + 0.5}{2} = 13.9 \qquad H = \frac{27.3 - 0.5}{2} = 13.4$$

The resulting coded value for the PPD concentration is, according to Eq. 4-19:

$$x^* = \frac{14.3 - 13.9}{13.4} = 0.03$$

B. On the basis of the coded value of a PPD concentration of 0.2 the original concentration is to be computed by means of the data in Table 4-13. For this we apply Eq. 4-22 and the sought concentration value is calculated as:

$$x = 0.2 \times 13.4 + 13.9 = 16.6 \text{ mM}$$

Factor effects versus regression parameters

Although we have considered factor-effect calculations and regression parameter estimation independently it is important to understand that both concepts are linked. More exactly, the following relationship holds:

$$\text{regression\_parameter} = \frac{\text{factor\_effect}}{\text{factor\_range}} \qquad (4\text{-}23)$$

Consider our enzyme example. Modeling the rate dependence by means of a polynomial that accounts for the same factors as in the screening design in Table 4-9 we will obtain the regression equation

$$y = 10.04 + 2.88\,\text{PPD} + 1.084\,\text{pH} + 0.4288\,\text{PPD}\times\text{pH} \qquad (4\text{-}24)$$

The insignificant regression parameters have been eliminated from this equation. If you now compare the factor effects D in Eq. 4-11 with the regression parameters you will find that the latter are half as large as the factor effects, i.e.,

$$D_{\text{PPD}} = 5.76$$

$$D_{\text{pH}} = 2.17$$

$$D_{\text{PPH·pH}} = 0.858$$

This is because the coded factor range is equal to 2 (from -1 to $+1$) for all three factors, and is therefore within rounding-off errors. In this example, the factor effects should be twices as large as the regression parameters.

Blocking of experiments

If a large number of experiments is to be carried out it will be difficult to run the experiments under identical conditions. During experimentation amounts of reagent might change or the activity of an enzyme might deteriorate. Often it will be necessary to interrupt experimentation during the night, and so important changes in the experimental conditions might result.

To reflect systematic changes in such situations the sequence of experiments has up to now been randomized. Strong systematic changes, however, will substantially increase the overall experimental error. Elimination of these systematic changes can be accounted for if the changes are taken as a discrete event and estimation of the time-dependent effects is confounded with estimation of unimportant interactions such as a three-factor interaction.

The experimental design is then divided into blocks.

Example 4-9: *Blocking of experiments*

Considering the 2^3-design of Table 4-4 the blocks could be designed as the half-fraction designs given in Table 4-14.

Table 4-14.
Full factorial 2^3-design arranged in two blocks

Experiment	Factors				Response
	x_1	x_2	x_3	$x^*(x_1 x_2 x_3)$	
Block 1:					
1	−1	−1	−1	+1	y_1
2	+1	+1	−1	+1	y_2
3	+1	−1	+1	+1	y_3
4	−1	+1	+1	+1	y_4
Block 2:					
5	+1	−1	−1	−1	y_5
6	−1	+1	−1	−1	y_6
7	−1	−1	+1	−1	y_7
8	+1	+1	+1	−1	y_8

Estimation of parameters for, e.g., the main effects would be performed by use of the following polynomial:

$$y = b_0 + b_1 x_1 + b_2 x_2 + b_3 x_3 + b^* x^* \qquad (4\text{-}25)$$

Because of the orthogonality of the experimental design, the changes between the blocks will not influence the estimation of the parameters b_0, b_1, b_2, and b_3. In addition, an averaged measure for the changes between the two blocks can be derived from the size of $2b^*$.

4.3 Sequential optimization: the simplex method

'Change and hope' is the name sometimes given to an optimization procedure where the individual factors are changed independently of each other. As long as the factors do not interact with each other the single-factor-at-a-time approach will succeed. In Fig. 4-16a this situation is explored for a response surface including curved factor effects. If one starts with variation of factor 2 keeping factor 1 at the coordinate value labeled (1) an optimal value will be found at the coordinate (2) for factor 2. In the next step factor 1 would be investigated with factor 2 constant at label (2) and the optimum would be found.

a.

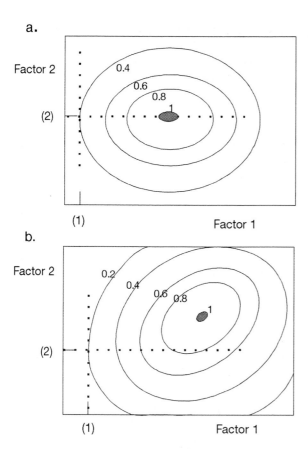

Factor 2

(2)

(1)   Factor 1

Fig. 4-16.
Sequential optimization with the
single-factor-at-a-time strategy
for response surfaces without
factor interactions (a) and with
interactions (b)

b.

Factor 2

(2)

(1)   Factor 1

However, if the factors interact the single-factor-at-a-time
approach does not guarantee that the optimum is reached. As
seen in Fig. 4-16b changing factor 2 will result in an optimal
coordinate at label (2) that would be fixed for changing of
factor 1. In this case the real optimum will never be found and
the result remains suboptimum. The reason is that the ridge in
Fig. 4-16b does not lie parallel to the factors axis. So changes
of factor 1 are not independent of factor 2. Instead both factors
interact and have to be considered simultaneously.

The most common sequential optimization method is based
on the simplex method of Nelder and Mead. A simplex is a
geometric figure having one more vertex than the number of
factors. Therefore a simplex in one dimension is a line, in two
dimensions a triangle, in three dimensions a tetrahedron, and in
multiple dimensions a hyper-tetrahedron.

Fixed-size simplex

To find the steepest path along a response surface by means
of the simplex method an algorithm has to be followed that
consists of designing an initial simplex, running the experi-

ments at the initial vertices and calculating the new vertex point by reflection of the vertex with the worst response. Movement with steps of fixed size is called the fixed-size simplex.

Fig. 4-17.
Fixed-size simplex according to Nelder and Mead along an unknown response surface

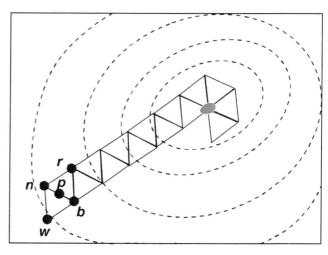

The algorithm works as follows (cf. Fig. 4-17):
- Generate the initial simplex according to the coded levels of the factors as given in Table 4-15.
- Run the experiments at the initial simplex coordinates.
- Decide from the responses which vertex represents the best response (vector **b**), the next-to-best response (**n**), and the worst response (**w**).
- Calculate the new experimental point from:

$$r = p + (p - w) \qquad (4\text{-}26)$$

where **p** is the centroid of the face remaining if the worst vertex **w** has been eliminated from the full simplex.

The centroid is calculated according to:

$$p = \frac{1}{n} \sum_{j \neq i}^{n} v_j \qquad (4\text{-}27)$$

where

n — number of factors
i — index of worst vertex to be eliminated
j — index of considered vertex

These steps are repeated until the simplex begins to rotate around the optimum or the response satisfies the experimenters' needs. The fixed step width of the fixed size simplex might lead to problems if the step width chosen is either too large or too small. In the first case, the optimum might be missed and in the latter case the number of experiments required becomes very large. These disadvantages can be circumvented if the step width is tunable, as with the variable-size simplex.

Table 4-15.
Choice of initial simplexes for up to nine variables coded in the interval between 0 and 1

Experiments	x_1	x_2	x_3	x_4	x_5	x_6	x_7	x_8	x_9
1	0								
2	1	0							
3	0.50	0.87	0						
4	0.50	0.29	0.82	0					
5	0.50	0.29	0.20	0.79	0				
6	0.50	0.29	0.20	0.16	0.78	0			
7	0.50	0.29	0.20	0.16	0.13	0.76	0		
8	0.50	0.29	0.20	0.16	0.13	0.11	0.76	0	
9	0.50	0.29	0.20	0.16	0.13	0.11	0.094	0.75	0
10	0.50	0.29	0.20	0.16	0.13	0.11	0.094	0.083	0.75

Variable-size simplex

With the variable-size simplex the step width is changed by expansion and contraction of the reflected vertices. The algorithm is modified as follows (cf. Fig. 4-18):

$$r = p + (p - w) \qquad (4\text{-}28)$$

(a) if r is better than b: expand the simplex

$$e = p + \alpha(p - w) \qquad (4\text{-}29)$$

with $\alpha > 1$, for example 1.5 for all directions or α is chosen different for each direction;

(b) if r lies between b and n: keep the simplex bnr

(c) if r is worse than n: contract the simplex according to:
1. r is worse than n but better than w, contract in the 'positive' direction:

$$c_+ = p + \beta(p - w) \qquad \text{with} \qquad 0 < \beta < 0.5 \qquad (4\text{-}30)$$

2. r is worse than w, contract in the 'negative' direction:

$$c_- = p + \beta(p - w) \qquad \text{with} \qquad 0 < \beta < 0.5 \qquad (4\text{-}31)$$

(d) At the experimental boundaries the simplex is reflected into the space of the experimental variables.
(e) Stop the simplex if the signal change is less than the experimental error or if the step width is less than a given threshold.

Fig. 4-18.
Variable-size simplex

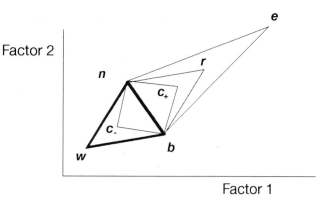

Factor 2

Factor 1

In practice, the simplex method is the most widely used experimental optimization algorithm. The main advantages are its simplicity, speed, and good convergence properties. Problems with the simplex method arise if multimodal response surfaces are investigated, i.e., if several local optima exist. In this case the simplex will climb the nearest local maximum or minimum and the global optimum might be missed. Mathematical theory provides more efficient optimization methods, such as the conjugate gradient method or Powell's method. These methods, however, are mainly used in locating optima of mathematical functions and are rarely used in experimental optimization.

Local optima are typical for optimization of selectivity in HPLC separations. This is caused by changes in the order of elution of peaks with different mobile phases.

Example 4-10: *Simplex optimization*

In this example the performance of the variable-size simplex is demonstrated for the enzyme determination based on the problem in Examples 4-4 and 4-7. For a fixed enzyme concentration of 13.6 mg L^{-1} ceruloplasmin (coded 0) the concentration of the substrate PPD and the pH value are sought for the maximum rate of the reaction, y. Since the simplex searches for a minimum the rate as the objective criterion has to be transformed. Here we use the difference from 100 min^{-1}, i.e. the objective criterion is $100 - y$.

The *initial simplex* is chosen according to the scheme in Table 4-15 in coded levels. The responses for the initial simplex are as follows:

vertex	coded variable		$100 - y$
	PPD	pH	
1	0.0	0.0	76.547
2	1.0	0.0	77.550
3	0.5	0.87	76.450

Fig. 4-19 demonstrates the initial simplex by use of bold lines.

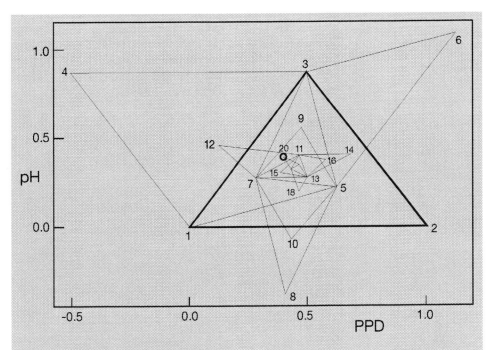

Fig. 4-19. Simplex search for optimum PPD concentration and pH value for the enzymatic determination of ceruloplasmin (cf. Example 4-7)

The best (minimum) response is found for vertex 3 (best response **b**), the next-to-best response is 1 (vertex **n**) and the worst response is found for vertex 2 (**w**). The last vertex is eliminated and a new (reflected) vertex 4 is calculated on the basis of the centroid (Eq. 4-27) according to Eq. 4-28:

centroid:

$$p = \frac{1}{2}\left[(0,0)+(0.5,0.87)\right] = (0.25,0.435)$$

reflection:

$$r = p+(p-w) = (0.25,0.435)+\left[(0.25,0.435)-(1,0)\right]$$
$$= (-0.5,0.87)$$

The new simplex now consists of the vertices 1, 3 and 4 which give the following responses:

| vertex | coded variable | | $100-y$ |
	PPD	pH	
1	0.0	0.0	76.547
3	0.5	0.87	76.450
4	−0.5	0.87	83.317

The vertex 4 (*r*) produces a worse response than the next-to-best vertex 1 and is even worse than *w*. Therefore, the simplex is contracted according to equation 4-31. At first the centroid is calculated without the worst vertex 4. Here the centroid is the same as in the first step:

centroid:

$$p = \frac{1}{2}[(0,0) + (0.5,0.87)] = (0.25,0.435)$$

contraction:

$$c_- = p - \beta(p-w) = (0.25,0.435)$$
$$- 0.5[(0.25,0.435) - (-0.5,0.87)] = (0.625,0.218)$$

The new simplex is 1, 3 and 5. The coordinates of this simplex and the measured responses are:

vertex	coded variable		$100 - y$
	PPD	pH	
1	0.0	0.0	76.547
3	0.5	0.87	76.450
5	0.625	0.218	75.131

After contraction the simplex is reflected to point 6 and again contracted to vertex 7 (cf. Fig. 4-19). If the calculation is continued the simplex moves in the direction of the optimum as given in the figure. After 20 iterations the following optimum is found in coded levels:

PPD = 0.46
pH = 0.316
$100 - y$ = 74.88

Thus, the maximum rate $y = 100 - 74.88 = 25.12$ min^{-1}. This optimum compares well with the results from the response surface study in Fig. 4-15.

4.4 General reading

Box, G. E. P., Hunter, W. G., Hunter, J. S., *Statistics for experimenters: An introduction to design, data analysis and model building*. New York: Wiley, 1987.
Box, G. E. P., Draper, N. R., *Empirical model-building and response surfaces*. New York: Wiley, 1987.

Deming, S. N., Morgan, S. L., *Experimental design: a chemometric approach.* Amsterdam: Elsevier, 1987.

Nelder, J. A., Mead, R., *A simplex method for function optimization.* Comput. J., 1965, **7**, 308.

Massart, D. L., Vandeginste, B. G. M., Deming, S. N., Michotte, Y., Kaufmann, L., *Chemometrics: a textbook.* Amsterdam: Elsevier Science Publishers, 1987.

Questions and Problems

1. Specify the following characteristics with a standard statistical software package
   • number of experiments
   • specific design
   • randomized run sequence
   • possible blocking
   • alias structures
   for two-level designs: 2^9, 2^{7-3}, 2^{10-5} as well as for central composite designs based on 3 and 6 factors.

2. Give definitions for the following terms: unimodal, multimodal, replication of experiments, factor level, coding of factors, local and global optimum, randomization, screening designs, factor effects.

3. Graphical inspection of response surfaces is restricted to 3D-plots. How do you plot response surfaces if more the two factors are included in the study?

4. How can one avoid factors and their parameter estimates being confounded?

5. How can one account for interactions of factors in a polynomial model?

6. What difficulties might arise if experimental factors are changed one-by-one?

7. Why do chemists run their experiments mostly by one-factor-at-a-time methods?

5 Pattern recognition and classification

Learning objectives

- To evaluate and interpret analytical data from complete chromatograms, spectra, depth profiles or electroanalytical records, from multidimensional detectors, and from samples for which concentrations of several chemical constituents or other properties have been measured
- To learn about methods for data preprocessing and for calculating distances and similarity measures
- To introduce grouping of analytical data based on unsupervised learning methods, i.e. projection methods and cluster analysis
- To handle multivariate data for which their class membership is known by means of supervised pattern recognition approaches

Modern analytical instrumentation generates a vast amount of data. A digitized spectrum in the IR-spectral range consists of about 2000 wavenumber points. In a GC-MS experiment, it is not difficult to generate in a single run 600000 items of data that amount to about 2.4 Mbytes of digital information. There are different methods of dealing with this extensive amount of information. One approach is ignorance. This means that quantitative analysis in spectroscopy is restricted to data evaluation at a single wavelength or the GC-MS trace is followed at a single mass unit. However, with the introduction of computers to the analytical laboratory most of the multidimensional data are stored in instrument computers or external data bases and so important information might be wasted if only small fractions of the data are evaluated. The extensive use nowadays of chemometrics enables the evaluation and interpretation of these data to be carried out efficiently as will be examined in this chapter.

The main objectives for application of multivariate methods in analytical chemistry are aimed at the grouping and classification of objects (samples, compounds, or materials) as well as at modeling relationships between different analytical data. Some typical examples include:

(a) *Grouping* or *clustering* of rock samples according to their similar elemental pattern, or of material samples in respect of comparable chemical composition and technological properties.

(b) *Classification* of samples, such as rocks, materials, or chemical compounds, by means of analytical data (spectrum, chromatogram, or elemental pattern) on the basis of known class membership of those objects.

Clustering and classification methods are summarized by the notion *pattern recognition*

(c) Calibration of a single chemical constituent by means of a full spectrum, or calibration of several components by means of mixture calibration techniques. In mathematical terms these are problems of *parameter estimation* where the parameters represent the calibration coefficients.

To understand the multidimensionality of analytical problems let us consider a practical example. In connection with the elucidation of a crime a hair recovered at the scene of the crime is to be assigned to a subject. Apart from the common morphological investigations the hair is analyzed for the elements copper, manganese, chlorine, bromine and iodine. For comparison the elemental content of three samples of hair from each of three suspects are available (Table 5-1).

Table 5-1.
Elemental content of hair of different suspects in ppm (parts per million)

Hair	Cu	Mn	Cl	Br	I
1	9.2	0.30	1730	12.0	3.6
2	12.4	0.39	930	50.0	2.3
3	7.2	0.32	2750	65.3	3.4
4	10.2	0.36	1500	3.4	5.3
5	10.1	0.50	1040	39.2	1.9
6	6.5	0.20	2490	90.0	4.6
7	5.6	0.29	2940	88.0	5.6
8	11.8	0.42	867	43.1	1.5
9	8.5	0.25	1620	5.2	6.2

Feature variables may be defined on different scales. The *nominal scale* characterizes qualitative equivalence, e.g. male and female. The *ordinal scale* describes ordering or ranking. The *interval scale* measures distances between values of the featues. The *ratio scale* enables also quotients between feature values to be evaluated.

The first check is whether the hair from the three suspects can be distinguished on the basis of their elemental patterns. For this chemometric methods for grouping of samples are needed. If grouping of the samples is feasible then the hair found must be assigned to a suspect on the basis of the construction of class models for the three subjects with subsequent classification of the unknown sample.

In general, the analytical data can be arranged as a data matrix X of n objects (rows) and p features (columns). The objects might be samples, molecules, materials, findings, or fertilizers. Typical features or variables of those objects will be elemental patterns, spectra, structural features, or physical properties. The $n*p$ data matrix X can be written as follows:

$$X = \begin{pmatrix} x_{11} & x_{12} & \cdots & x_{1p} \\ x_{21} & x_{22} & \cdots & x_{2p} \\ \vdots & & & \\ x_{n1} & x_{n2} & \cdots & x_{np} \end{pmatrix} \quad (5\text{-}1)$$

For some problems these data are divided column-wise into dependent and independent variables, e.g., for calibrating concentrations on spectra. The dependent variables are then renamed, for example, by use of the character y.

A *class* comprises a collection of objects which have similar features. A *pattern* of an object is its collection of characteristic features. For multivariate data evaluation, not all objects and features are necessarily used. On the other hand, some of the available data cannot be used as they are reported. Therefore, pretreatment of data is a prerequisite for efficient multivariate data analysis.

5.1 Preprocessing of data

Missing data, centering, and scaling

In the first step, the data have to be reviewed for completeness. *Missing data* do not hinder mathematical analysis. Of course, missing data should not be replaced by zeros. Instead the vacancies should be filled up either by the column/row mean or, in the worst case, by generating a random number in the range of the considered column or row. *Features* can be removed from the data set if they are highly correlated with other features, or if they are redundant or constant.

To eliminate a constant offset the data can be translated along the coordinate origin. The common procedure is *mean centering*, where each variable, x_{ik}, is centered by subtracting the column mean, $\bar{x}_k$, according to:

$$x_{ik}^* = x_{ik} - \bar{x}_k \qquad (5\text{-}2)$$

where i is the row index, k the column index, and $\bar{x}_k$ is the column mean calculated from:

$$\bar{x}_k = \frac{1}{n}\sum_{i=1}^{n} x_{ik} \qquad (5\text{-}3)$$

Coded data are obtained by multiplying, dividing, adding and/or subtracting a constant in order to convert the original data into more convenient values.

Very often the features represent quite different properties of a sample or of an object, so that the metric might differ substantially from column to column. This might imply different absolute values of the variables as well as different variable ranges (variances). Both types of distortion will affect most of the statistically based multivariate methods. These differences can be eliminated by *scaling* the data to similar ranges and variances. Two scaling methods are important, those that scale the data by range or by standard deviation (autoscaling):

Range scaling:

$$x_{ik}^* = \frac{x_{ik} - x_k(\text{min})}{x_k(\text{max}) - x_k(\text{min})} \qquad 0 \le x_{ik}^* \le 1 \qquad (5\text{-}4)$$

Autoscaling:

$$x_{ik}^* = \frac{x_{ik} - \bar{x}_k}{s_k} \qquad (5\text{-}5)$$

$$\text{where } s_k = \sqrt{\frac{\sum\limits_{i=1}^{n}(x_{ik} - \bar{x}_k)}{n-1}} \qquad (5\text{-}6)$$

and n is the number of objects. Autoscaling reveals data with zero mean and unit variance. The length of the vectors is scaled to $\sqrt{n-1}$.

Normalization to length one

In some cases, normalization of a data vector to length one is an important preprocessing procedure:

$$x_{ik}^* = \frac{x_{ik}}{\|x_k\|} \qquad (5\text{-}7)$$

$$\text{where } \|x_k\| = \sqrt{x_{1k}^2 + x_{2k}^2 + \dots x_{nk}^2} \qquad (5\text{-}8)$$

Usually the autoscaling method is the method of choice for scaling data. Fig. 5-12 gives a graphical illustration of centering and autoscaling.

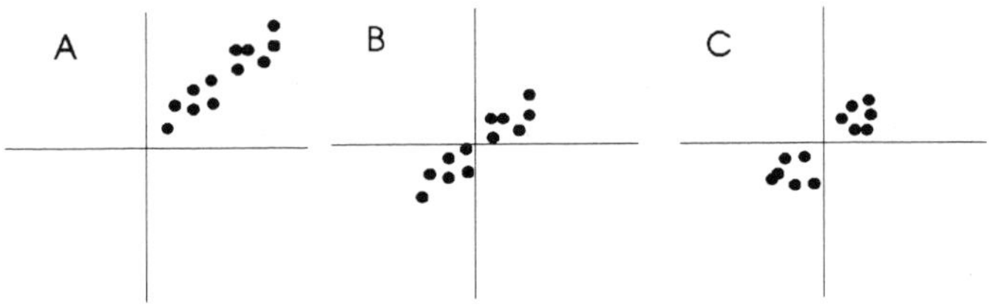

Fig. 5-1.
Demonstration of translation and scaling procedures: the original data in A are centered in B and autoscaled in C. Notice that the autoscaling reduces the between-groups distance in the direction of greatest within-groups scatter, and increases it in perpendicular direction in the sense of sphericization of groups.

Variance-covariance matrix and correlation matrix

Transformation of the original data to a new coordinate system is another possibility for data pretreatment. The methods are based on principal component analysis or factor analysis.

The first step for these transformations is the formation of a data matrix that is derived from the original data matrix and that reflects the relationships among the data. Such derived data matrices are the variance-covariance matrix and the correlation matrix.

Variance-covariance matrix

The variance-covariance matrix or simply covariance matrix is computed from the data matrix, X, in Eq. 5-1 from the variances of all p variables by use of Eq. 5-9 and their covariances according to Eq. 5-10.

$$s_{ij}^2 = \frac{1}{n-1} \sum_{i=1}^{n} (x_{ij} - \bar{x}_j)^2 \quad \text{for} \quad j = 1 \dots p \qquad (5\text{-}9)$$

$$\text{cov}(j,k) = \frac{1}{n-1} \sum_{i=1}^{n} (x_{ij} - \bar{x}_j)(x_{ik} - \bar{x}_k) \qquad (5\text{-}10)$$

$$j, k = 1 \dots p; \quad j \neq k$$

For the variance-covariance matrix Eq. 5-11 is valid:

$$C = \begin{pmatrix} s_{11}^2 & \text{cov}(1,2) & \dots & \text{cov}(1,p) \\ \text{cov}(2,1) & s_{22}^2 & & \text{cov}(2,p) \\ \vdots & \vdots & & \vdots \\ \text{cov}(p,1) & \text{cov}(p,2) & \dots & s_{pp}^2 \end{pmatrix} \qquad (5\text{-}11)$$

As a result a symmetric matrix is obtained. The covariance matrix is used in cases where the metrics of the variables are comparable. If the metrics are widely different, e.g., for simultaneous data treatment of main and trace constituents, the variables are scaled. If the variables are autoscaled (cf. Eq. 5-5) the correlation matrix given in Eq. 5-14 results automatically.

At this point only one possibility for computing the variance-covariance matrix has been introduced. Another one is described in Sec. 5.2.1 in connection with factor analysis.

Correlation matrix

To calculate the correlation matrix the correlation coefficients are required according to:

$$r_{jk} = \frac{\text{cov}(j,k)}{s_j s_k} = \frac{\sum\limits_{i=1}^{n}(x_{ij} - \bar{x}_j)(x_{ik} - \bar{x}_k)}{\left[\sum\limits_{i=1}^{n}(x_{ij} - \bar{x}_j)^2 \sum\limits_{i=1}^{n}(x_{ik} - \bar{x}_k)^2\right]^{1/2}} \quad j \neq k$$

$$(5\text{-}12)$$

By analogy with Eq. 5-9 the standard deviations s_j and s_k are calculated from:

$$s_j = \sqrt{\frac{\sum\limits_{i=1}^{n}(x_{ij} - \bar{x}_j)^2}{n-1}} \qquad s_k = \sqrt{\frac{\sum\limits_{i=1}^{n}(x_{ik} - \bar{x}_k)^2}{n-1}} \qquad (5\text{-}13)$$

The correlation matrix reads as follows:

$$R = \begin{pmatrix} 1 & r_{12} & \cdots & r_{1p} \\ r_{12} & 1 & & r_{2p} \\ \vdots & & & \vdots \\ r_{1p} & r_{2p} & \cdots & 1 \end{pmatrix} \qquad (5\text{-}14)$$

Computations of the covariance and correlation matrix are prerequisites for the application of factorial methods.

5.2 Unsupervised methods

Grouping of analytical data is possible either by means of clustering methods or by projecting the high dimensional data on to lower dimensional space. Since there is no supervisor in the sense of known membership of objects to classes these methods are performed in an unsupervised manner.

5.2.1 Factorial methods

These methods are aimed at projecting the original data set from high dimensional space on to a line, a plane, or a 3D-coordinate system. Perhaps the best way would be to have a mathematical procedure that enables you to sit in front of the computer screen pursuing the rotation of the data in all possible directions and stopping this process when the best projection, i.e., optimal clustering of data groups, has been found. In fact, such methods of projection pursuit already exist in statistics and are tested within the field of chemometrics.

The *dispersion matrix* describes the scatter of multivariate data around the mean. For centered data the dispersion matrix equals X^TX.

At present, data projection is performed mainly by methods called principal component analysis (PCA), factor analysis (FA), singular value decomposition (SVD), eigenvector projection, or rank annihilation. The different methods are linked to different areas of science. They also differ mathematically in the way the projection is computed, i.e. which dispersion matrix is the basis for data decomposition, which assumptions are valid, and whether the method is based on eigenvector analysis, singular value decomposition, or on other iterative schemes.

PCA – principal component analysis

An explanation of projection methods is based here on PCA in comparison with SVD. Similar methods, such as factor analysis, can be understood later in the section.

The key idea of PCA is to approximate the original matrix X by a product of two small matrices – the score and loading matrices – according to:

$$X = T L^T \qquad (5\text{-}15)$$

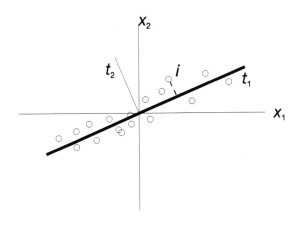

where X is the original data matrix consisting of n rows (objects) and p columns (features); T is the scores matrix with n rows and d columns (number of principal components); L is the loading matrix with d columns and p rows; and T is the transpose of a matrix.

In other words, the projection of X down on to a d-dimensional *subspace* by means of the projection matrix L^T gives the object coordinates in this plane, T. The columns in T are the score vectors and the rows in P are called loading vectors. Both vectors are orthogonal, i.e., $p_i^T p_j = 0$ and $t_i^T t_j = 0$, for $i \neq j$.

In contrast with factor analysis the data are reconstructed such that new uncorrelated variables are obtained. The principal components are determined on the basis of the maximum variance criterion (Fig. 5-2). Each subsequent principal component describes a maximum of variance, that is not modeled by the former components. According to this most of the variance of the data is contained in the first principal component. In the

Fig. 5-2.
Projection of a swarm of objects from the original two dimensions on to one dimension, i.e., the score vector t_1, according to the criterion of maximum variance

125

second component there is more information than in the third, etc. Finally as many principal components are computed as are needed to explain a preset percentage of the variance (cf. Eq. 5-22).

The principal components, PCs for short, can be considered as projections of the original data matrix, X, on to the scores, T. For this Eq. 5-15 must be converted to the scores on the left side by:

$$T = XL \tag{5-16}$$

The new coordinates are linear combinations of the original variables. For example, the elements of the first principal component read as:

$$t_{11} = x_{11}l_{11} + x_{12}l_{21} + ... + x_{1p}l_{p1} \tag{5-17}$$

$$t_{21} = x_{21}l_{11} + x_{22}l_{21} + ... + x_{2p}l_{p1}$$

$$\vdots$$

$$t_{n1} = x_{n1}l_{11} + x_{n2}l_{21} + ... + x_{np}l_{p1}$$

Because a large fraction of the variance can usually be described by means of one, two or three principal components, the data can be visualized by plotting the principal components against each other (cf. Example 5-2).

Principal components in PCA or common factors in factor analysis are sometimes called *latent variables*.

The simplest method used for PCA in analytical chemistry is the iterative NIPALS (**n**onlinear **i**terative **p**artial **l**east **s**quares) algorithm explained in Example 5-1. More powerful methods are based on matrix diagonalization, such as singular value decomposition, or bidiagonalization, such as the partial least squares method.

Example 5-1: *Iterative principal component analysis by the NIPALS algorithm*

Determine principal components for the following data matrix on the basis of the iterative NIPALS algorithm:

$$X = \begin{pmatrix} 2 & 1 \\ 3 & 2 \\ 4 & 3 \end{pmatrix}$$

0. Scaling by the mean and normalizing to length one:

$$X = \begin{pmatrix} -\dfrac{1}{\sqrt{2}} & -\dfrac{1}{\sqrt{2}} \\ 0 & 0 \\ \dfrac{1}{\sqrt{2}} & \dfrac{1}{\sqrt{2}} \end{pmatrix}$$

1. Estimation of the loading vector l^T. Usually the first row of the X-matrix is used:

$$l^T = \left(-\dfrac{1}{\sqrt{2}} \quad -\dfrac{1}{\sqrt{2}} \right)$$

2. Computation of the new score vector t:

$$t = Xl = \begin{pmatrix} -\dfrac{1}{\sqrt{2}} & -\dfrac{1}{\sqrt{2}} \\ 0 & 0 \\ \dfrac{1}{\sqrt{2}} & \dfrac{1}{\sqrt{2}} \end{pmatrix} \begin{pmatrix} -\dfrac{1}{\sqrt{2}} \\ -\dfrac{1}{\sqrt{2}} \end{pmatrix} = \begin{pmatrix} 1 \\ 0 \\ -1 \end{pmatrix} \qquad (5\text{-}18)$$

Comparison of the new and old t-vector. If the deviations of the elements of the two vectors are within a given threshold of 10^{-z}, e.g., $z = 5$, then continue at step 5, otherwise go to step 3.

3. Compute new loadings l^T:

$$l^T = t^T X = (1 \quad 0 \quad -1) \begin{pmatrix} -\dfrac{1}{\sqrt{2}} & -\dfrac{1}{\sqrt{2}} \\ 0 & 0 \\ \dfrac{1}{\sqrt{2}} & \dfrac{1}{\sqrt{2}} \end{pmatrix} \qquad (5\text{-}19)$$

$$= \left(-\dfrac{2}{\sqrt{2}} \quad -\dfrac{2}{\sqrt{2}} \right)$$

Normalize the loading vector to length 1:

$$l^T = \dfrac{l^T}{\|l^T\|} = \left(-\dfrac{1}{\sqrt{2}} \quad -\dfrac{1}{\sqrt{2}} \right) \qquad (5\text{-}20)$$

4. Repeat from step 2 if the number of iterations does not exceed a predefined threshold, e.g. 100; otherwise go to step 5.

5. Determine the matrix of residuals:

$$E = X - tl^T = \begin{pmatrix} -\dfrac{1}{\sqrt{2}} & -\dfrac{1}{\sqrt{2}} \\ 0 & 0 \\ \dfrac{1}{\sqrt{2}} & \dfrac{1}{\sqrt{2}} \end{pmatrix}$$

(5-21)

$$- \begin{pmatrix} 1 \\ 0 \\ -1 \end{pmatrix} \left(-\dfrac{1}{\sqrt{2}} \quad -\dfrac{1}{\sqrt{2}} \right) = \begin{pmatrix} 0 & 0 \\ 0 & 0 \\ 0 & 0 \end{pmatrix}$$

If the number of principal components is equal to the number of previously fixed principal components or the number of cross-validated components then go to step 7. Otherwise continue at 6.

6. Use the residual matrix E as the new X-Matrix and compute additional principal components t and loadings l^T by means of step 1.

7. As a result, the matrix X is represented by a principal component model according to Eq. 5-15, i.e.,

$$X = TL^T = \begin{pmatrix} 1 \\ 0 \\ -1 \end{pmatrix} \left(-\dfrac{1}{\sqrt{2}} \quad -\dfrac{1}{\sqrt{2}} \right)$$

The actual two-dimensional data can be described by use of just one principal component.

With real data, more principal components are necessary. Therefore, there are more columns of scores in the T-matrix and more rows in the L^T-matrix representing the loadings.

Estimating the number of principal components

The use of all principal components after decomposition of the data matrix is usually not justified. For example the number of pure components must be separated from the noise components in a spectro-chromatogram.

To decide on the the number of components in a PCA several heuristic and statistical criteria exist:
• percentage of explained variance
• eigenvalue-one criterion
• Scree-test
• cross validation

The percentage of *explained variance* is applied in the sense of a heuristic criterion. It can be used if enough experience is gained by analyzing similar data sets. The fraction of explained (cumulative) variance, s_e^2, is calculated from the ratio of the sum of d important eigenvalues and the sum of all p eigenvalues by use of:

$$s_e^2 = \frac{\sum_{i=1}^{d} \lambda_i}{\sum_{i=1}^{p} \lambda_i} \qquad (5\text{-}22)$$

If all possible principal components are used in the model 100% of the variance is explained. Usually a fixed percentage of explained variance is specified, e.g. 90%. Eq. 5-22 is then multiplied by 100.

In our example of hair data in Table 5-1 90.7% of the data variance can still be explained by use of two principal components (Table 5-2).

Table 5-2.
Eigenvalues and explained variances for the hair data in Table 5-1

Component	Eigenvalue λ	Explained variance %	Cumulative variance %
1	3.352	67.05	67.05
2	1.182	23.65	90.70
3	0.285	5.70	96.40
4	0.135	2.70	99.10
5	0.045	0.90	100.00

The *eigenvalue-one criterion* is based on the fact that the average eigenvalue of autoscaled data is just one. This is because for standardized data the sum of all eigenvalues of the correlation matrix is exactly the number of all features p (cf. Eq. 5-15). Only components with eigenvalues greater than 1 are considered important. According to this criterion the eigenvalues of the hair data in Table 5-2 reveal two significant principal components.

The *Scree-test* is based on the phenomenon that the residual variance levels off when the proper number of principal components is obtained. Visually the residuals or more often the eigenvalues are plotted against the number of components in a Scree-plot. The component number is then derived from the leveling-off of this dependence. Fig. 5-3 depicts the Scree-plot for the hair data. The slope can be seen to drop between the second and third components.

The fourth method for deciding on the number of principal components is *cross validation*. In the simplest case, every object of the X-matrix is removed from the data set once, and a model is computed with the remaining data. Then the removed data are predicted by means of the PCA model and the sum of the

Fig. 5-3.
Scree-plot for the principal component model of the hair
data of Table 5-1

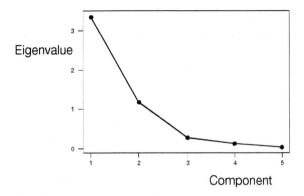

Eigenvalue

Component

square roots of the residuals over all removed objects is calculated. The number of significant principal components is
obtained from the minimum residual error.

For large data sets, the leave-one-out method can be replaced by leaving out groups of objects. Other criteria for deciding on the number of principal components or factors are introduced below.

Graphical interpretation of principal components

Interpretation of the results of principal component analysis
is usually carried out by visualization of the component scores
and loadings. Sometimes the data can be interpreted from a
single component. Commercial software provides two- or
three-dimensional plot facilities. The following example demonstrates the general procedure.

Example 5-2: *Principal component analysis*

The data of elemental composition of hair in Table 5-1
must be investigated by PCA. First the grouping of the
samples must be recognized and second the importance of
different elements for discrimination between the groups
must be discussed.

Fig. 5-4.
Principal component scores
for the hair data in Table 5-1

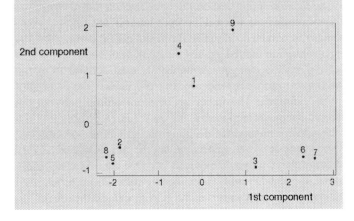

2nd component

1st component

PCA on the basis of the correlation matrix of the data provides the results given in Fig. 5-4 for the scores and Fig. 5-5 for the loadings. Because the preliminary tests revealed only two significant principal components (cf. Table 5-1 and Fig. 5-3) plots of the first two principal components are sufficient.

The score plot depicts the linear projection of objects representing the main part of the total variance of the data. As can be seen, there are three clusters with three objects (hair) each. These objects belong to the hair from different subjects.

Correlation and importance of feature variables must be decided from plots of the PC loadings. Fig. 5-5 shows the loading plot of the first two components for the hair data of Table 5-1.

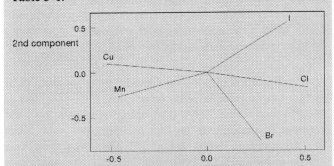

Fig. 5.5.
Principal component loadings of the hair data in Table 5-1

The loading plot provides the projection of the features on to the principal components. From this plot, information about the *correlation* of feature variables can be deduced. The correlation of features is described by the cosine of the angle between the loading vectors. The smaller the angle the higher is the correlation between features. Uncorrelated features are orthogonal to each other. If variables are highly correlated then it is sufficient to measure only one of the correlated variables.

The size of the loadings in relation to the considered principal component is a measure of the *importance* of a feature for the PC model. Loadings in the origin of the coordinate system represent unimportant features.

In our hair data example (Fig. 5-5) there is greater correlation for the elements Cu and Mn than for the halogens Br, Cl, and I. For Cu and I anticorrelation is observed. All elements are important for describing the first principal component. The second component is mainly characterized by the elements I and Br.

A joint interpretation of scores and loadings is possible if the loadings are properly scaled and superimposed on the score plots. The so-called *biplot* for the current example is given in Fig. 5-6. The discriminating ability of variables can be deduced from the loading direction. In our example the features Cu, Mn, I, and Cl separate the object clusters into the groups (2,8,5) and (3,6,7)

Karhungen-Loeve expansion is synonymous with principal component analysis.

131

Fig. 5-6
Biplot for the simultaneous
characterization of the scores
and loadings of two principal
components of the hair data in
Table 5-1

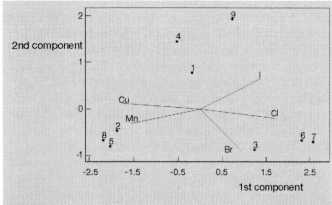

whereas the feature Br separates the left cluster (2,5,8) from
the rest of the objects. The proximity of objects to a loading
vector reflects the importance of that variable for building
the principal component model.

The computation of principal components in Example (5-2)
was based on singular value decomposition. If the same task is
solved by use of the NIPALS algorithm (Example 5-1), the
projection of data on to the first two principal components
gives the result presented in Fig. 5-7. Apart from those two
methods many additional possibilities exist for decomposition
of the original data matrix. As a further method we will learn
about factor analysis later in this section.

Fig. 5-7.
Biplot for the simultaneous
characterization of the scores
and loadings of two principal
components of the hair data in
Table 5-1 computed on the basis
of the NIPALS algorithm

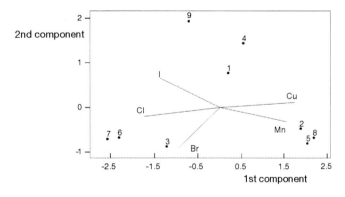

To understand the relationship between principal component
analysis and factor analysis it is useful to outline the method of
singular value decomposition in more detail here.

SVD – singular value decomposition

SVD is based on the decomposition of a symmetric matrix,
e.g. the correlation matrix, into a threefold diagonal matrix and
the diagonalization of these by means of the so-called QR
algorithm. Details on the SVD algorithm are not important here.

To understand the relationships between principal component, factor and eigenvalue analysis, however, it is useful to discuss the principle of SVD in matrix representation.

SVD decomposes the data matrix, X, in Eq. 5-1 into the matrices U, W and V as follows:

$$X = UWV^{\mathrm{T}} \tag{5-23}$$

$$
\begin{pmatrix}
x_{11} & x_{12} & \cdots & x_{1p} \\
x_{21} & x_{22} & & x_{2p} \\
\vdots & & & \\
x_{n1} & x_{n2} & \cdots & x_{np}
\end{pmatrix}
=
\begin{pmatrix}
u_{11} & u_{12} & \cdot & u_{1d} \\
u_{21} & u_{22} & & u_{2d} \\
\vdots & & & \\
u_{n1} & u_{n2} & \cdot & u_{nd}
\end{pmatrix}
\begin{pmatrix}
w_{11} & 0 & \cdot & 0 \\
0 & w_{22} & & 0 \\
\vdots & & & \\
0 & 0 & \cdot & w_{dd}
\end{pmatrix}
\begin{pmatrix}
v_{11} & v_{21} & \cdots & v_{p1} \\
v_{12} & v_{22} & & v_{p2} \\
\vdots & & & \\
v_{1d} & v_{2d} & & v_{pd}
\end{pmatrix}
$$
$$\tag{5-24}$$

The matrix U contains the same column vectors as does the scores matrix T in Eq. 5-15 but normalized to length one; W is the diagonal matrix containing the square roots of the eigenvalues or singular values. For symmetric matrices $(n = p)$ the singular values are identical to the square roots of the eigenvalues, i.e.

$$w_{ii} = \sqrt{\lambda_{ii}} \tag{5-25}$$

The dimension of the matrix of singular values corresponds to the number of columns in the scores matrix. If small singular values are not neglected, i.e. not set to zero, the dimension of W is equal to the number of features, i.e. $d = p$.

The matrix V^{T} is identical to matrix L^{T} in Eq. 5-15. U and V are also denoted as left and right vectors of singular values, respectively.

FA – factor analysis

The aim of factor analysis (FA) is to express the features by use of a small number of *common* factors. For each factor a property is assigned that cannot be observed directly. For example, if air is analyzed in a city a common factor could be traffic. FA was originally developed for explaining psychological theories, e.g. for description of factors such as 'intelligence' or 'memory'.

To transform the abstract factors determined in the first step into *interpretable* factors rotation methods are applied. If definite target vectors can be assumed to be contained in the data, e.g. a spectrum under a spectro-chromatogram, the rotation of data is performed by using a target. This technique is known as target-transform factor analysis (TTFA, cf. Example 5-6).

In PCA the objects are usually associated with samples or more generally with cases and features with the properties of those cases. In contrast to this in FA the properties are arranged as the objects in the rows and the samples as features in the columns.

The general strategy of factor analysis is again the decomposition of the data matrix, X, in Eq. 5-1 into two smaller matrices F and L:

$$X = FL^T + E \tag{5-26}$$

$$
\begin{pmatrix}
x_{11} & x_{12} & \cdots & x_{1p} \\
x_{21} & x_{22} & & x_{2p} \\
\vdots & & & \vdots \\
x_{n1} & x_{n2} & \cdots & x_{np}
\end{pmatrix}
=
\begin{pmatrix}
f_{11} & f_{12} & \cdot & f_{1d} \\
f_{21} & f_{22} & & f_{2d} \\
\vdots & & & \vdots \\
f_{n1} & f_{n2} & \cdot & f_{nd}
\end{pmatrix}
\begin{pmatrix}
l_{11} & l_{21} & \cdot & l_{p1} \\
l_{12} & l_{22} & & l_{p2} \\
\vdots & & & \vdots \\
l_{1d} & l_{2d} & \cdot & l_{pd}
\end{pmatrix}
+
\begin{pmatrix}
e_{11} & e_{12} & \cdots & e_{1p} \\
e_{21} & e_{22} & & e_{2p} \\
\vdots & & & \vdots \\
e_{n1} & e_{n2} & & e_{np}
\end{pmatrix}
$$

$$\tag{5-27}$$

The scores matrix, F, contains the values of the propertices of the d causal factors. The columns of the matrix L characterise the fraction of the *loading* related to the considered factor. The matrix E consists of the $p - d$ remaining or *specific* factors. Those factors cannot be related to the common factors sought. Frequently the specific factors represent random interferences, e.g. noise factors.

Q- and R-analysis

The starting point in FA is, as in PCA, the covariance or correlation matrix. Until now only one possibility of computing this dispersion matrix has been exploited (Eqs. 5-11 and 5-14). In principle the dispersion matrices can be calculated by multiplication of the data matrix by its transpose or by relating the transposed matrix to the original.

Consider the calculation of the *variance-covariance matrix* of the scaled data of the X-matrix:

$$C_Q = X^T X \quad \text{or} \quad C_R = XX^T \tag{5-28}$$

where X is an $n \times p$ matrix (cf. Eq. 5-1), and Q, R are labels for the so-called Q- and R-analysis modes, respectively.

Computation of the covariance matrix based on Q-mode has been tacitly used in Eqs. 5-9 to 5-11. A matrix, C_Q, of dimensions $p \times p$ is obtained. The covariance matrix for the R-mode, C_R, is determined after transposing X. Its dimensions are $n \times n$.

For the correlation matrix we obtain for the two modes:

$$R_Q = (XV_Q)^T (XV_Q) \quad \text{or} \quad R_R = (V_R X)(V_R X)^T \quad (5\text{-}29)$$

where

$$v_Q(ij) = \cfrac{1}{\cfrac{1}{p-1}\sqrt{\sum_{i=1}^{p}(x_{ij} - \bar{x}_j)^2}}$$

$$(5\text{-}30)$$

$$v_R(ij) = \cfrac{1}{\cfrac{1}{n-1}\sqrt{\sum_{i=1}^{n}(x_{ij} - \bar{x}_j)^2}}$$

Evaluation based on the Q- or R-mode are also needed in connection with other multivariate methods. In general the technique is called R-analysis, when the relationship among p features determined by n observations is of interest. Q-analysis examines the relationship among n objects characterized by p variables. For the most frequently used approach in factor analysis the interpretation of the loadings can be denoted R-analysis and, as seen below, the evaluation of the scores as Q-analysis.

Communalities and reduced correlation matrix

As mentioned above we distinguish in factor analysis between common and specific factors. The criterion for this distinction is based on their loadings that are clearly different from zero. A common factor is found if at least two of its loadings are distinctly different from zero. For a specific factor it is true that only one of the loadings l_{1k}... l_{pk} is clearly distinguished from zero. The subdivision of loadings, L', into d common factors and $p-d$ specific factors can be expressed as the sum of the loadings of common factors, L'', and of specific factors, L''':

$$L' = L'' + L''' = \begin{pmatrix} l_{11} & l_{21} & \cdots & l_{1d} & 0 & \cdots & 0 \\ l_{21} & l_{22} & & l_{2d} & 0 & & 0 \\ & & & & & \vdots & \\ l_{p1} & l_{p2} & \cdots & l_{pd} & 0 & \cdots & 0 \end{pmatrix} + \begin{pmatrix} 0 & \cdots & 0 & l_{1d+1} & 0 & \cdots & 0 \\ 0 & & 0 & 0 & l_{2d+2} & & 0 \\ \vdots & & & & & & \vdots \\ 0 & \cdots & 0 & 0 & 0 & \cdots & l_{pp} \end{pmatrix} \quad (5\text{-}31)$$

For orthogonal factors we obtain:

$$R = LL^T + KK^T \quad (5\text{-}32)$$

where

$$
L = \begin{pmatrix} l_{11} & l_{21} & \cdots & l_{1d} \\ l_{21} & l_{22} & & l_{2d} \\ \vdots & & & \vdots \\ l_{p1} & l_{p2} & \cdots & l_{pd} \end{pmatrix} \qquad K = \begin{pmatrix} l_{1d+1} & 0 & \cdots & 0 \\ 0 & l_{2d+2} & & 0 \\ \vdots & & & \vdots \\ 0 & 0 & \cdots & l_{pp} \end{pmatrix} \qquad (5\text{-}33)
$$

The squared elements of matrix K are exactly the specific variances.

$$
K^2 = KK^T = \begin{pmatrix} l_{1d+1}^2 & 0 & \cdots & 0 \\ 0 & l_{2d+2}^2 & & 0 \\ \vdots & & & \\ 0 & 0 & \cdots & l_{pp}^2 \end{pmatrix} = \begin{pmatrix} \sigma_1^2 & 0 & \cdots & 0 \\ 0 & \sigma_2^2 & & 0 \\ \vdots & & & \vdots \\ 0 & 0 & \cdots & \sigma_p^2 \end{pmatrix}
$$

$$(5\text{-}34)$$

These variances correspond to the fractions of the standardized variance, that can be explained by the common factors.

On the basis of the loading matrix, L, for orthogonal factors a reduced correlation matrix of the following form results:

$$
R' = LL^T = R - KK^T = \begin{pmatrix} h_1^2 & r_{12} & \cdots & r_{1p} \\ r_{12} & h_2^2 & & r_{2p} \\ \vdots & & & \\ r_{1p} & r_{2p} & & h_p^2 \end{pmatrix} \qquad (5\text{-}35)
$$

where $h_j^2 = 1 - \sigma_j^2 = l_{j1}^2 + l_{j2}^2 + \ldots + l_{jd}^2$ for $j = 1 \ldots p$

In Eq. 5-35 the quantities h_j^2 are the so-called *communalities*. They reflect which fraction of the variance of the jth standardized feature is explained by common factors. The communalities are either empirically known or they can be estimated from random samples.

For correlated (skewed) factors the communalities are derived from the multiple correlation coefficients according to:

$$
h_j^2 = 1 - \frac{1}{r_{jj}} \qquad \text{for} \qquad j = 1 \ldots p \qquad (5\text{-}36)
$$

The first problem of a factor analysis consists in estimation of the loading matrix, L. We will consider only two methods here, i.e. principal component analysis and principal factor analysis.

Estimation of factor loadings

Factor analysis in a more narrow sense means procedures where the *reduced* correlation matrix, R', (Eq. 5-35) is reproduced. For that the communalities are required. The reproduction of the original correlation matrix R (Eq. 5-14), as done with PCA, is not a factor analysis in a restrictive sense. However, because PCA has been discussed already above, we will use it here as basis for performing a factor analysis.

Principal component analysis. The loading matrix has to reproduce the correlation matrix R, i.e.:

$$R = LL^{\mathrm{T}} \tag{5-37}$$

The assumptions about specific factors as made in factor analysis in the more narrow sense are not made in this context. The factorial model reads, in the terminology of factor analysis (cf. Eq. 5-15):

$$X = FL^{\mathrm{T}} \tag{5-38}$$

As discussed above PCA is usually not carried out to derive interpretable components. But complicated relationships should be reduced to simple ones by projecting the data from multi-dimensional space to two or three dimensions.

For this the principal axes are translated and rotated without changing the distances of the features relative to each other. The pairwise perpendicularly arranged coordinates remain orthogonal. Mathematically this transformation can be dealt with by solving an eigenvalue problem. The loading vector l_k of the kth component corresponds to the normalized eigenvector of the related kth largest eigenvalue λ_k of the empirical correlation matrix R, i.e.

$$l_k^{\mathrm{T}} l_k = 1 \tag{5-39}$$

For better understanding eigenvalue analysis is described here in more detail.

Eigenvector analysis

For a symmetric, real matrix, R, an eigenvector v is obtained from:

$$Rv = v\lambda \tag{5-40}$$

where λ is an unknown scalar – the *eigenvalue*. The eigenvector must be determined such that the vector Rv is proportional to v. For this, Eq. 5-40 is rewritten as:

$$Rv - v\lambda = 0 \quad \text{or} \quad (R - \lambda I)v = 0 \tag{5-41}$$

where I is the identity matrix and the vector, v, is orthogonal to all of the row vectors of matrix $(R - \lambda I)$. The equation obtained is equivalent to a set of d where d is the rank of R.

Unless v is the null vector, Eq. 5-41 is valid only if:

$$R - \lambda I = 0 \qquad (5\text{-}42)$$

A solution to this set of equations exists only if the determinant on the left side of Eq. 5-42 is zero:

$$|R - \lambda I| = 0 \qquad (5\text{-}43)$$

Computing the determinant reveals a polynomial in λ of degree d. Then the roots, λ_i with $i = 1 \dots d$, of those equations must be found. There will be an associated vector v_i, such that:

$$Rv_i - v_i \lambda_i = 0 \qquad (5\text{-}44)$$

or in matrix notation:

$$RV = V\Lambda \qquad (5\text{-}45)$$

The matrix V is a square and orthogonal matrix of the eigenvectors. The different ways of computing the dispersion matrix by the Q- or R-analysis techniques lead to different sets of eigenvalues, as we will see below in a comparison with singular value decomposition.

Example 5-3: *Eigenvalue determination*

As an example of an eigenvalue analysis we use the following data matrix consisting of three rows and two columns:

$$\begin{pmatrix} 0.9 & 4.1 \\ 1.9 & 2.9 \\ 2.9 & 2.1 \end{pmatrix}$$

The autoscaled form of this matrix is:

$$X = \begin{pmatrix} -1 & 1.059 \\ 0 & -0.132 \\ 1 & -0.927 \end{pmatrix} \qquad (5\text{-}46)$$

For determination of the eigenvalues Eqs. 5-40 to 5-45 are applied. First the correlation matrix is calculated:

$$R = \begin{pmatrix} 1 & -0.993 \\ -0.993 & 1 \end{pmatrix} \qquad (5\text{-}47)$$

Insertion of the eigenvalues, λ, into Eq. 5-43 leads to:

$$|R - \lambda I| = \begin{vmatrix} 1-\lambda & -0.993 \\ -0.993 & 1-\lambda \end{vmatrix} = 0 \qquad (5\text{-}48)$$

Calculation of the determinant results in the characteristic polynomial for the root λ. In our example a squared equation is obtained:

$$(1-\lambda)^2 - 1.993 = 0 \qquad (5\text{-}49)$$

Solution of Eq. 5-49 reveals the two eigenvalues $\lambda_1 = 1.993$ and $\lambda_2 = 0.115$.

For determination of the eigenvectors according to Eq. 5-44 we insert both eigenvalues into Eq. 5-48.

Calculation of the first eigenvector:

$$\begin{pmatrix} 1-1.993 & -0.993 \\ -0.993 & 1-1.993 \end{pmatrix} \begin{pmatrix} v_{11} \\ v_{21} \end{pmatrix} = \begin{pmatrix} 0 \\ 0 \end{pmatrix} \qquad (5\text{-}50)$$

Evaluation of Eq. 5-50 results in the following equation system with two unknowns:

$$\begin{aligned} -0.993v_{11} - 0.993v_{21} &= 0 \\ -0.993v_{11} - 0.993v_{21} &= 0 \end{aligned} \qquad (5\text{-}51)$$

Further simplification leads to:

$$v_{11} = -v_{21} \qquad (5\text{-}52)$$

For the eigenvectors in Eq. 5-44 there is an infinite number of solutions. For example, assume $v_{11} = 1$, then as a consequence $v_{21} = -1$. Usually the eigenvectors are normalized to length one, i.e.

$$v_{11}^2 + v_{21}^2 = 1 \qquad (5\text{-}53)$$

The predefined values of the eigenvectors are then divided by their size. i.e. $\sqrt{1^2 + 1^2} = \sqrt{2}$. For the first eigenvector we obtain:

$$\begin{pmatrix} v_{11} \\ v_{21} \end{pmatrix} = \begin{pmatrix} \dfrac{1}{\sqrt{2}} \\ -\dfrac{1}{\sqrt{2}} \end{pmatrix} = \begin{pmatrix} 0.707 \\ -0.707 \end{pmatrix} \qquad (5\text{-}54)$$

Analogous calculation of the second eigenvector gives:

$$\begin{pmatrix} 1-0.115 & -0.993 \\ -0.993 & 1-0.115 \end{pmatrix} \begin{pmatrix} v_{12} \\ v_{22} \end{pmatrix} = \begin{pmatrix} 0 \\ 0 \end{pmatrix} \tag{5-55}$$

$$\begin{aligned} 0.885v_{12} & -0.993v_{22} & = 0 \\ -0.993v_{12} & +0.885v_{22} & = 0 \end{aligned} \tag{5-56}$$

$$v_{12} = v_{22} \tag{5-57}$$

$$v_{12}^2 + v_{22}^2 = 1 \tag{5-58}$$

The second eigenvector is then:

$$\begin{pmatrix} v_{12} \\ v_{22} \end{pmatrix} = \begin{pmatrix} \dfrac{1}{\sqrt{2}} \\ \dfrac{1}{\sqrt{2}} \end{pmatrix} = \begin{pmatrix} 0.707 \\ 0.707 \end{pmatrix} \tag{5-59}$$

The solution for the matrix in equation Eq. 5-45 is then:

$$RV = V\Lambda = \begin{pmatrix} 1 & -0.993 \\ -0.993 & 1 \end{pmatrix} \begin{pmatrix} 0.707 & 0.707 \\ -0.707 & 0.707 \end{pmatrix} = \begin{pmatrix} 0.707 & 0.707 \\ -0.707 & 0.707 \end{pmatrix} \begin{pmatrix} 1.993 & 0 \\ 0 & 0.115 \end{pmatrix} \tag{5-60}$$

The matrix of eigenvectors V here directly provides the loading matrix L in Eq. 5-38. In the next step the matrix of scores F must be determined. A general explanation of this is given below. In Example 5-4 the score matrix will be computed for the data matrix X in Eq. 5-46.

Principal factor analysis. PCA can be slightly modified to be applied as a factor analytical method in its genuine sense, if the specific variances are included. For this, the communalities $h_1^2 \ldots h_p^2$ of the p features (cf. Eq. 5-36) are estimated from the correlation matrix R and the reduced correlation matrix R' is reproduced according to Eq. 5-35 from the loadings:

$$R' = LL^{\mathrm{T}} \tag{5-61}$$

The reduced correlation matrix is subsequently subjected to principal component analysis: the eigenvalues are determined and normalized to length one, as explained in Example 5-3. The significant eigenvectors then determine the loading matrix L. This approach is termed principal factor analysis.

More powerful factor analytical methods are the centroid method, the maximum-likelihood method and canonical factor analysis. With those methods the loading matrix is estimated iteratively. Details can be obtained from the available software and the statistical literature [5-1].

Determination of the scores

Apart from the loadings for judging the objects it is also important to know the scores. Unique scores can be determined only on the basis of a complete principal component analysis, since the loading matrix is orthonormal, which means:

$$L^T L = I \qquad (5\text{-}62)$$

The scores matrix can be directly estimated from the model for the standardized data matrix, X, since:

$$X = FL^T \qquad (5\text{-}38)$$

in more detail $X = FL^T L = FI = F$ $\qquad (5\text{-}63)$

For factor analysis in its strict sense no unique scores matrix can be given, because the loading matrix only reproduces the reduced correlation matrix of the features. Therefore the score matrix F must be estimated. The estimation methods must be taken from the specialized literature.

Example 5-4: *Determination of the factor score matrix*

For the data in Example 5-3 the matrix of factor scores F is to be computed.
To obtain that score matrix the values for the standardized matrix X (Eq. 5-46) and the loading matrix L or V computed in Example 5-3, Eq. 5-60, are inserted into Eq. 5-63:

$$F = XL = \begin{pmatrix} -1 & 1.059 \\ 0 & -0.132 \\ 1 & -0.927 \end{pmatrix} \begin{pmatrix} 0.707 & 0.707 \\ -0.707 & 0.707 \end{pmatrix} = \begin{pmatrix} -1.455 & 0.0422 \\ 0.095 & -0.0937 \\ 1.363 & 0.0515 \end{pmatrix}$$

$$(5\text{-}64)$$

Decomposition of the standardized data matrix in Eq. 5-46 by PCA results in the following factor analytical model:

$$X = \begin{pmatrix} -1 & 1.059 \\ 0 & -0.132 \\ 1 & -0.927 \end{pmatrix} = FL^T = \begin{pmatrix} -1.455 & 0.0422 \\ 0.095 & -0.0937 \\ 1.363 & 0.0515 \end{pmatrix} \begin{pmatrix} 0.707 & -0.707 \\ 0.707 & 0.707 \end{pmatrix}$$

$$(5\text{-}65)$$

The matrix U represents the eigenvector of $X^T X$ (left eigenvector) and the matrix V is the eigenvector of XX^T (right eigenvector).

If the factor analysis is performed on the basis of principal component analysis a direct relationship to singular value decomposition (SVD) can be derived (cf. Eq. 5-2).

Denote the matrix, R_Q, of the eigenvectors that were determined from the correlation matrix by Q-analysis by U and the eigenvector matrix by R-analysis, R_R, by V, then decomposition of the matrix X by means of SVD gives:

$$X = U\Lambda^{1/2}V^T \tag{5-66}$$

Depending on the mode, i.e. whether Q- or R-analysis has been used for computation of Λ, the dimensions of matrix Λ are either $n \times n$ or $p \times p$, respectively.

The factor analytical solutions after R- and Q-analysis are interrelated as follows:

$$X = L_Q F_Q \quad \text{or} \quad X = F_R L_R \tag{5-67}$$

where

$$X = \overbrace{U}^{L_Q} \overbrace{\Lambda^{1/2}V^T}^{F_Q} \qquad X = \overbrace{U\Lambda^{1/2}}^{F_R} \overbrace{V^T}^{L_R} \tag{5-68}$$

The direction of the solution is less dependent on the mode, Q- or R-analysis, but is directed by the scaling procedures applied. The scaling may be based on scaling the columns or rows or on the scaling of both.

Example 5-5: *Factor analysis by R-mode*

Eigenvalue analysis in Example 5-3 was carried out by R-analysis, i.e. the matrix of scores F_R (Eq. 5-64) is formed according to Eq. 5-68 by the product of matrix U from SVD and the matrix of the eigenvalues (Eq. 5-49):

$$F_R = U\Lambda^{1/2} = \begin{pmatrix} -0.729 & 0.367 \\ 0.476 & -0.815 \\ 0.683 & 0.448 \end{pmatrix} \begin{pmatrix} 1.993 & 0 \\ 0 & 0.115 \end{pmatrix}$$
$$= \begin{pmatrix} -1.455 & 0.0422 \\ 0.095 & -0.0937 \\ 1.363 & 0.0515 \end{pmatrix} \tag{5-69}$$

Compare the result with that in Eq. 5-64.

Determination of the number of significant factors

To decide on the number of common factors some criteria have already been introduced in connection with PCA, such as the eigenvalue-one-test or the Scree test. Those criteria can also be used in factor analysis, if the determination of loadings is performed by principal component analysis.

For factor analysis in its genuine sense additional criteria are used for rank analysis, i.e. for determination of the number of significant factors. Most frequently an empirical indicator function, *IND*, introduced by Malinowski [5-2] is used. It is computed from the real error, *RE*, or the residual standard deviation *RSD* as follows:

$$IND = \frac{RE}{(p-d)^2} = \frac{RSD}{(p-d)^2} = \frac{\left(\sum\limits_{k=d+1}^{p} \frac{\lambda_k}{n(p-d)} \right)^{1/2}}{(p-d)^2} \qquad (5\text{-}70)$$

where p is the number of features, d the number of common factors, n the number of objects and λ_k the kth eigenvalue. The indicator function has a minimum at the most probable number of factors.

Rotation methods

An optimal loading matrix is obtained by rotation of factors. One distinguishes between orthogonal and oblique (correlated) rotations. For an orthogonal rotation the coordinate system is rotated. The aim is that the new coordinate axis cuts the group of points in an optimal way. This can be often better achieved by an oblique rotation. If the data can be described by an orthogonal rotation in an optimal way, then also an oblique method will lead to coordinate axes that are perpendicular to each other.

For an orthogonal rotation of the loading matrix L a rotated loading matrix L_{rot} is obtained by muliplication of that matrix with a transformation matrix T:

$$L_{rot} = LT \qquad (5\text{-}71)$$

For an oblique rotation by multiplication with a transformation matrix only a matrix of so-called factor structure, L_{fst} is obtained. This matrix contains correlation of common factors and features:

$$L_{fst} = LT \qquad (5\text{-}72)$$

The rotated loading matrix is only obtained after multiplication of L_{fst} by the inverse correlation matrix of factors:

$$L_{rot} = L_{fst} R_F^{-1} \qquad (5\text{-}73)$$

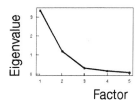

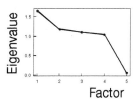

Eigenvalues without (above) and with *rotation* after the varimax-criterion (below) for the hair data in Table 5-1.

As examples of orthogonal and oblique factor rotations the varimax, quartimax and oblimax criteria will be considered below.

The *varimax-criterion* serves the purpose of an orthogonal rotation, where the variance of the squared loadings within a common *factor* is maximized. As a result as many common factors should be retained as are described by as few features (variables) as possible. Large eigenvalues and loadings are increased, but small ones are reduced. The uniform weighting of the variables is guaranteed by scaling the loadings by means of the communalities (cf. Eq. 5-35). The varimax-criterion is calculated from:

$$\max\left[\mathrm{V}\right] = p\sum_{j=1}^{p}\sum_{k=1}^{d}\sigma_{jk}^{2} - \sum_{k=1}^{d}\left(\sum_{j=1}^{p}\sigma_{jk}^{2}\right)^{2} \tag{5-74}$$

where p is the number of features and d the number of factors.

In contrast the *quartimax-criterion* maximizes the variance of the (squared) loadings of a *variable*. The aim is to express each variable by as few common factors as possible. This kind of rotation leads to an increase of large loadings and a decrease of small loadings of each variable. However, it might happen that a single common factor emerges from the rotation. The quartimax-criterion is obtained from the communalities as follows:

$$\max\left[\mathrm{Q}\right] = \sum_{k=1}^{d}\sum_{j=1}^{p}\sigma_{jk}^{4} \tag{5-75}$$

For the *oblimax-criterion* a function similar to that known for the quartimax-criterion is maximized, this is termed kurtosis function:

$$\max\left[\mathrm{K}\right] = \frac{\sum_{k=1}^{d}\sum_{j=1}^{p}\sigma_{jk}^{4}}{\left[\sum_{k=1}^{d}\sum_{j=1}^{p}\sigma_{jk}^{4}\right]^{2}} \tag{5-76}$$

More criteria for rotation of factors are given in, e.g., reference [5-3].

Target-rotation: target-transformation factor analysis

A special kind of rotation matches the abstract factor loadings to known patterns. For example, it is tested on the basis of whether, in a spectro-chromatogram, a hypothetical spectrum can be found under an incompletely resolved chromatographic peak or whether a compound pattern can be detected in the exhaust gas of an emitter. The hypothetical spectrum or the elemental pattern is termed the *target*, L^{*}. The containment of a target in a data matrix is evaluated by computing the target-transformation matrix, T.

The starting point is the model of factor analysis in Eq. 5-26:

$$X = FL^T$$

The abstract loadings, L, are tested against a hypothetical loading vector, L^*, i.e. against the targets. This relationship is given on the basis of the transformation matrix T by:

$$LT = L^* \tag{5-77}$$

Computation of the transformation matrix is feasible by multiple linear regression analysis (cf. Sec. 6.2):

$$T = \left(L^T L\right)^{-1} L^T L^* \tag{5-78}$$

The fit of the target vectors in matrix L^* and the predicted vectors in matrix L, based on the transformation matrix according to Eq. 5-77, are usually estimated from the average relative deviation as follows:

$$relative\ deviation = \frac{\sum\limits_{j=1}^{p}\left|l_j^* - \hat{l}_j\right|}{\sum\limits_{j=1}^{p}\left|l_j^*\right|} \tag{5-79}$$

Let us consider the different steps of a target transformation factor analysis for example data from the combination of liquid chromatography with UV-spectroscopy.

Example 5-6: *Target-transformation factor analysis*

In a HPLC chromatogram polycyclic aromatic hydrocarbons (PAH) are expected under an incompletely resolved peak. Since detection was based on a diode-array detector, the observed peak can be evaluated at several wavelengths. The spectrochromatogram for five wavelengths and seven retention times is shown in Fig. (5-8). The corresponding data are given in Table 5-3.

Table 5-3.
Absorbances in milliabsorbance units for the spectro-chromatogram in Fig. 5-8

Retention time, min	Wavelength, nm 245	265	285	305	325
6.4	7.81	4.83	4.367	0.944	1.775
6.5	84.33	52.69	56.100	12.890	20.730
6.6	161.58	99.30	108.430	26.920	39.026
6.7	173.33	77.89	97.260	39.368	28.670
6.8	274.70	63.92	82.160	47.150	20.060
6.9	218.92	36.95	39.820	25.580	10.490
7.0	79.04	12.07	10.580	6.536	3.230

Fig. 5-8.
Simplified spectro-chromatogram for an incompletely resolved peak in HPLC with diode-array detection. Absorbance is given in milliabsorbance units

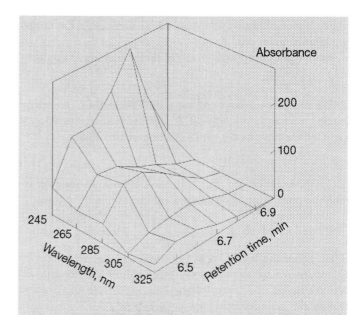

In the first step we perform factor analysis on the basis of the data for the spectro-chromatogram in Table 5-3. The result is three significant common factors with the following scores and loadings:

$$X = FL^{\mathrm{T}} = \begin{pmatrix} 1.431 & -0.375 & 0.839 \\ 0.183 & -0.889 & -0.266 \\ -1.077 & -1.302 & -0.977 \\ -0.881 & -0.186 & 1.509 \\ -0.868 & 1.387 & 0.480 \\ 0.110 & 1.175 & -1.297 \\ 1.102 & 0.191 & -0.286 \end{pmatrix} \begin{pmatrix} -0.779 & -0.958 & -0.979 & -0.892 & -0.905 \\ 0.606 & -0.282 & -0.195 & 0.417 & -0.423 \\ -0.282 & -0.050 & 0.058 & 0.177 & -0.047 \end{pmatrix} \quad (5\text{-}80)$$

To rotate the loading matrix the target spectra of the hypothetical compounds benzo[k]fluoranthene (B[k]F), benzo-[b]fluoranthene (B[b]F), perylene and anthracene are tested. These spectra are illustrated in Fig. 5-9 for the considered wavelengths.

On the basis of the computed loadings (Eq. 5-80) we obtain for the transformation matrix T the following matrix equation:

$$LT = L^* = \begin{pmatrix} -0.779 & 0.606 & -0.282 & 1 \\ -0.958 & -0.282 & -0.049 & 1 \\ -0.979 & -0.195 & 0.058 & 1 \\ -0.892 & 0.417 & 0.177 & 1 \\ -0.905 & -0.423 & -0.047 & 1 \end{pmatrix} \quad T = \begin{pmatrix} 111.2 & 112.6 & 282.1 & 280.0 \\ 38.2 & 87.2 & 76.4 & 2.25 \\ 52.5 & 69.4 & 12.2 & 1.0 \\ 110.6 & 33.2 & 5.1 & 1.3 \\ 14.7 & 25.0 & 6.9 & 5.5 \end{pmatrix} \quad (5\text{-}81)$$

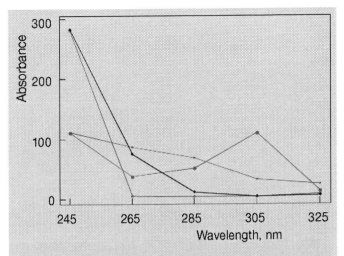

Fig. 5-9.
Simplified UV-spectra
of the compounds
benzo[*k*]fluoranthene (●),
benzo[*b*]fluoranthene (■),
perylene (◆) and anthracene (▲)
as the targets

Note that the features (wavelengths) in the loading matrix
L are arranged in the rows. The target spectra are given in
the hypothetical loading matrix L^* column-wise in the order
of the compounds B[*k*]F, B[*b*]F, perylene and anthracene.
The values correspond to the absorbances shown in Fig. 5-9.
Since in our case the spectra do not need to be centralized,
we have to add a column of ones in the loading matrix. Then
the intercepts for the spectra on the ordinate can be accounted for in the calculation of the transformation matrix.

The transformation matrix is computed by means of multiple linear regression according to Eq. 5-78:

$$T = \begin{pmatrix} -141.1 & -782.5 & -589.5 & 769.1 \\ 112.5 & 134.5 & 242.7 & 111.2 \\ 14.57 & -423.5 & -888.3 & -249.7 \\ -64.69 & -645.9 & -465.2 & 748.4 \end{pmatrix} \qquad (5\text{-}82)$$

Prediction of the target spectra by multiplication of the
loading matrix with the transformation matrix reveals:

$$\hat{L} = LT = \begin{pmatrix} 111.2 & 112.6 & 282.4 & 256.2 \\ 38.1 & 87.0 & 75.3 & -7.5 \\ 52.4 & 69.6 & 13.4 & -40.7 \\ 110.6 & 33.1 & 4.7 & 64.9 \\ 14.7 & 25.0 & 7.1 & 17.2 \end{pmatrix} \qquad (5\text{-}83)$$

Comparison of the predicted matrix in Eq. 5-83 with the
hypothetical loading matrix in Eq. 5-81 demonstrates, that
the spectra from the first three columns of the compounds
B[*k*]F, B[*b*]F and perylene fit quite well. The compounds
are readily identified under the incompletely resolved peak.
But for the spectrum of anthracene (fourth column in the
loading matrix) an obvious deviation is observed between

Fig. 5-10.
Comparion of the target spectrum
of anthracene (●) with the
predicted spectrum (■) in Eq. 5-83

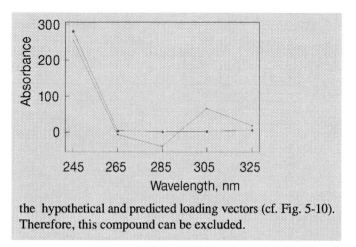

the hypothetical and predicted loading vectors (cf. Fig. 5-10).
Therefore, this compound can be excluded.

Apart from transformations of spectra, elution profiles in chromatography, depth profiles in materials analysis or elemental patterns in environmental analysis may also be interesting targets.

Evolving factor analysis (EVA)

This method is a further development of factor analysis, whereby intrinsically ordered data can be investigated. Multi-wavelength detection of the dependence on time of the elution of compounds in a chromatogram is a typical example. Also the spectroscopic investigation of the dependence on pH-value of simultaneous equilibria can be carried out by EVA.

For forward decomposition of the original data matrix the eigenvalues in EVA are computed by stepwise addition of objects. The evolving eigenvalues reveal the existence of individual components (cf. Fig. 5-11). In this way, for example, the intervals for the elution of compounds in a chromatogram can be evaluated.

Features

X

Fig. 5-11.
Schematic representation
of the dependence of evolving
eigenvalues, λ, on intrinsically
ordered data

5.2.2 Cluster analysis

The second strategy of unsupervised learning is based on cluster analysis. With this method the objects are aggregated stepwise according to the similarity of their features. As a result hierarchically or nonhierarchically ordered clusters are formed. In order to describe the similarity of objects we need to learn about appropriate similarity measures.

Distance and similarity measures

To decide on the similarity of objects the *distance measures* commonly applied in pattern recognition are used. The shorter the distance between objects the more similar they are.

A general distance measure is the Minkowski distance or the L_p-metric:

$$d_{ij} = \left[\sum_{k=1}^{K} \left| x_{ik} - x_{jk} \right|^p \right]^{1/p} \tag{5-84}$$

where

K – number of variables
i, j – indices for objects i and j

In most cases the Euclidean distance is applied, for which $p = 2$. For example, the Euclidean distance between two objects 1 and 2 is:

$$d_{12} = \left[(x_{11} - x_{21})^2 + (x_{12} - x_{22})^2 \right]^{1/2} \tag{5-85}$$

In the narrow sense cluster analysis should not be confused with classification methods, where unknown objects are assigned to existing classes. Cluster analyses belong to the methods of unsupervised learning or unsupervised pattern recognition.

A *cluster* describes a group of objects, that are more similar to each other than to objects outside the group. A single object or the centroid (•) serves as the *seed* of a cluster.

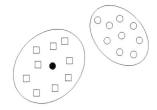

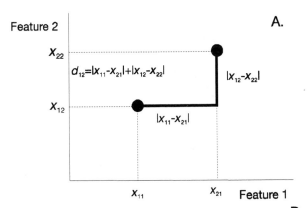

A.

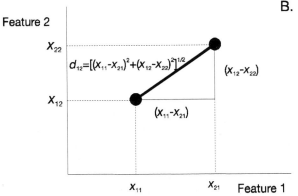

B.

Fig. 5-12.
City-block distance (A) and Euclidean distance (B) for two features

When $p = 1$ the so-called Manhattan or city-block distance is obtained. This distance refers to travel around a corner, i.e.:

$$d_{ij} = \sum_{k=1}^{K} \left| x_{ik} - x_{jk} \right| \qquad (5\text{-}86)$$

Fig. 5-12 demonstrates the city-block and the Euclidean distances graphically.

A disadvantage of measures based on the L_p-metric is their dependence on the metrics used. Therefore, scaling of data is frequently unavoidable if these distance measures are to be applied.

A distance measure based on the standard deviation of a variable j, s_j, is the *Pearson-distance*:

$$d_{ij} = \sqrt{\frac{\sum_{k=1}^{K} \left(x_{ik} - x_{jk} \right)^2}{s_j^2}}$$

A measure that accounts for the different scales of variables and, in addition, for their correlations is the Mahalanobis distance. This invariant measure is calculated by use of the following formula:

$$D_{ij}^2 = (\boldsymbol{x}_i - \boldsymbol{x}_j)^{\mathrm{T}} \, \boldsymbol{C}^{-1} \, (\boldsymbol{x}_i - \boldsymbol{x}_j) \qquad (5\text{-}87)$$

where $\boldsymbol{C}$ — covariance matrix (Eq. 5-11)
$\boldsymbol{x}_i, \boldsymbol{x}_j$ — column vectors for objects i and j, resp.

Scaling of data is not necessary if the Mahalanobis distance is used. In addition, with this measure distortion arising from correlations of features or feature groups is avoided. In constrast, if the Euclidean distance is to be applied for two highly correlated variables, these variables would be used as two independent features although they provide identical information.

Complementary to distance measures are *similarity measures*. For example, the similarity measure, S_{ij}, on the basis of the Minkowski distance is.:

$$S_{ij} = 1 - \frac{d_{ij}}{d_{ij}(\text{max})} \qquad (5\text{-}88)$$

where $d_{ij}(\text{max})$ represents the maximum distance of objects found in the data. Completely similar objects have a similarity measure of $S_{ij} = 1$. For completely unsimilar objects a value $S_{ij} = 1$ is expected.

Hierarchical cluster analysis

Hierarchical cluster analysis is deduced from *Taxonomy*, where biological species are ordered with respect to phenomenological similarities.

One possibility for clustering objects is their hierarchical aggregation. Here the objects are combined according to their distances from or similarities to each other. We distinguish agglomerative and divisive procedures. Divisive cluster formation is based on splitting the whole set of objects into individual clusters. With the more frequently used agglomerative clustering one starts with single objects and merges them to larger object groups.

150

In order to better understand the different steps of cluster analyses, hierarchical (agglomerative) clustering is demonstrated here for a data set from clinical analysis.

Example 5-7: *Cluster analysis (hierarchical)*

The grouping of patients' serum samples is to be investigated on the basis of cluster analysis. Features are the concentrations of calcium and phosphate analyzed in the serum samples. The values are given in Table 5-4.

Table 5-4.
Concentrations of calcium and phosphate in six blood serum samples

Object (serum sample)	Features Calcium, mg $(100\ mL)^{-1}$	Phosphate, mg $(100\ mL)^{-1}$
1	8.0	5.5
2	8.25	5.75
3	8.7	6.3
4	10.0	3.0
5	10.25	4.0
6	9.75	3.5

In the first step we calculate the distance matrix for all the data on the basis of one of the distance measures (Eqs. 5-84 to 5-87). Here we use the Euclidean distance. As an example the distance between objects 1 and 2 is evaluated by taking into account the feature values from Table 5-4:

$$d_{12} = \left[(8 - 8.25)^2 + (5.5 - 5.75)^2 \right]^{1/2} = 0.354$$

Calculation of the other object distances is carried out by analogy, i.e. every object is to be compared with the remaining objects. The distance between one and the same object is zero:

Object	1	2	3	4	5	6
1	0					
2	0.354	0				
3	1.063	0.711	0			
4	3.201	3.260	3.347	0		
5	2.704	2.658	2.774	1.031	0	
6	2.658	2.704	2.990	0.559	0.707	0

Reduction of the distance matrix is performed by aggregation of objects. Objects with the shortest distances between them are aggregated first. In this example the method of *weighted average linkage* is used for aggregation, where the objects are combined by averaging the calculated distances. The following steps demonstrate the building of clusters.

Clusters of objects with 2 or 3 features can be graphically represented. The visual clustering of blood sera in Table 5-4 gives:

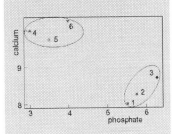

In multidimensional space cluster analysis becomes obligatory.

151

First reduced matrix: The shortest distance in the distance matrix is between the objects 1 and 2, i.e. $d_{12} = 0.354$. The two objects are aggregated to a new object 1* and their new distance is set to zero. The distance matrix is recomputed by averaging the individual distances as follows:

$$d_{1*3} = \frac{d_{13} + d_{23}}{2} = \frac{1.063 + 0.711}{2} = 1.774$$

$$d_{1*4} = \frac{d_{14} + d_{24}}{2} = \frac{3.202 + 3.260}{2} = 3.231$$

$$d_{1*5} = \frac{d_{15} + d_{25}}{2} = \frac{2.704 + 2.658}{2} = 2.681$$

$$d_{1*6} = \frac{d_{16} + d_{26}}{2} = \frac{2.658 + 2.704}{2} = 2.681$$

We obtain for the reduced matrix:

Object	1*	3	4	5	6
1*	0				
3	1.774	0			
4	3.231	3.347	0		
5	2.681	2.774	1.031	0	
6	2.681	2.990	0.559	0.707	0

Second reduced matrix: The shortest distance is observed here for the objects 4 and 6 with $d_{46} = 0.559$. The two objects 4 and 6 form a new object 4*, the distance of which is again set to zero. The combination of the d-values of that row reveals the actual distance matrix:

$$d_{54*} = \frac{d_{54} + d_{56}}{2} = \frac{1.031 + 0.707}{2} = 0.869$$

$$d_{4*3} = \frac{d_{43} + d_{63}}{2} = \frac{3.347 + 2.990}{2} = 3.169$$

$$d_{4*1*} = \frac{d_{41*} + d_{61*}}{2} = \frac{3.231 + 2.681}{2} = 2.956$$

Object	1*	3	4*	5
1*	0			
3	1.774	0		
4*	2.956	3.169	0	
5	2.681	2.774	0.869	0

Third reduced matrix: The shortest distance is now $d_{54*} = 0.869$. A new object 5* is defined and a new distance matrix arises:

$$d_{1*5*} = \frac{d_{51*} + d_{4*1*}}{2} = \frac{2.681 + 2.956}{2} = 2.819$$

$$d_{35*} = \frac{d_{4*3} + d_{53}}{2} = \frac{3.169 + 2.774}{2} = 2.972$$

Object	1*	3	5*
1*	0		
3	1.774	0	
5*	2.819	2.972	0

Fourth reduced matrix: Here the objects 1* and 3 are aggregated with $d_{1*3} = 1.774$ to object 3*. Finally the distance to the remaining object 5* is evaluated:

$$d_{3*5*} = \frac{d_{5*1*} + d_{5*3}}{2} = \frac{2.819 + 2.972}{2} = 2.895$$

Object	3*	5*
3*	0	
5*	2.895	0

Graphically the distances between the clusters can be depicted as a *dendrogram* (cf. Fig. 5-13).

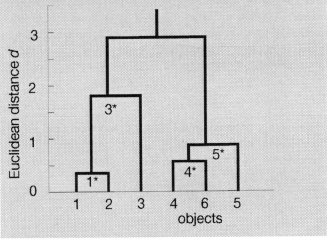

Fig. 5-13.
Dendrogram for the clinical analytical data from Table 5-4

To decide on the number of clusters different criteria can be exploited. Very often the number of clusters is known. In the given example on clinical data the number of clusters might be predefined by a given number of diseases. In some cases the number of clusters can be deduced from a predetermined distance measure or difference between the clusters.

In Example 5-7 aggregation of clusters was carried out by the weighted average linkage method. In general, the distance

153

to a new object or cluster k (labeled in the example by a star) is computed by calculating the average distance from the objects A and B to object i:

Weighted average linkage

$$d_{ki} = \frac{d_{Ai} + d_{Bi}}{2} \qquad (5\text{-}89)$$

The sizes of the clusters and their weights are assumed to be equal.

Several other formulas exist for construction of the distance matrix. The most important ones are considered now for the case of aggregating two clusters.

Single linkage

Here the shortest distance between the opposite clusters is calculated, i.e.:

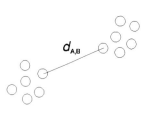

$$d_{ki} = \frac{d_{Ai} + d_{Bi}}{2} - \frac{|d_{Ai} - d_{Bi}|}{2} = \min(d_{Ai}, d_{Bi}) \qquad (5\text{-}90)$$

As a result clusters are formed that are loosely bound. The clusters are often linearly elongated in contrast to the usual spherical clusters. This chaining is caused by the fusion of single objects to a cluster. The procedure is related to the K-nearest neighbor method.

Complete linkage

This method is based on the largest distance between objects of the opposite clusters to be aggregated:

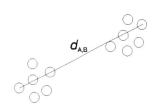

$$d_{ki} = \frac{d_{Ai} + d_{Bi}}{2} + \frac{|d_{Ai} - d_{Bi}|}{2} = \max(d_{Ai}, d_{Bi}) \qquad (5\text{-}91)$$

In general well separated, small, compact spherical clusters are formed.

Unweighted average linkage

With this method the number of objects in a cluster is used for weighting the cluster distances:

$$d_{ki} = \frac{n_A}{n} d_{Ai} + \frac{n_B}{n} d_{Bi} \quad \text{with} \quad n = n_A + n_B \qquad (5\text{-}92)$$

where n_A and n_B are the numbers of objects in clusters A and B, respectively. No deformation of the cluster is observed. To some extent small clusters consisting of outliers might arise.

Centroid linkage

Here the centroid calculated as the average of a cluster is applied as the basis for aggregation without distorting the cluster space:

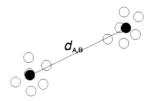

$$d_{ki} = \frac{n_{Ai}}{n} d_{Ai} + \frac{n_B}{n} d_{Bi} - \frac{n_A n_B}{n^2} d_{AB} \qquad (5\text{-}93)$$

Median linkage

For determination of the centroid the median can also be used:

$$d_{ki} = \frac{d_{Ai}}{2} + \frac{d_{Bi}}{2} - \frac{d_{AB}}{4} \qquad (5\text{-}94)$$

An advantage is that the importance of a small cluster is preserved after aggregation with a large one.

Ward's method

With the Ward-method the clusters are aggregated such that a minimum increase in the within-group error sum of squares results:

$$d_{ki} = \frac{n_A + n_i}{n + n_i} d_{Ai} + \frac{n_B + n_i}{n + n_i} d_{Bi} - \frac{n_i}{n + n_i} d_{AB} \qquad (5\text{-}95)$$

The error sum of squares is defined as the sum of the squared deviations of each from the centroid of its own cluster. This aggregation method leads to well-structured dendrograms. It is probably the method most frequently used.

A more general procedure can be derived by applying a distance formula as introduced by Lance and Williams:

$$d_{ki} = \alpha_A d_{Ai} + \alpha_B d_{Bi} + \beta d_{AB} + \gamma |d_{Ai} - d_{Bi}| \qquad (5\text{-}96)$$

The dependence on the choice of the parameters α, β, γ results in limiting cases which correspond to the above mentioned agglomeration methods (Table 5-5).

Dendrograms for cluster analysis of the hair data in Table 5-1 based on the Euclidean distance and the single linkage (A) or Ward-method (B).

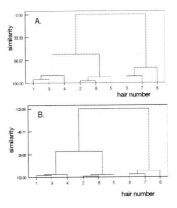

Table 5-5.

Parameters for hierarchical cluster analysis by means of the general distance formula after Lance and Williams in Eq. (5-96)

Method	α_A	α_B	β	γ
unweighted average	$n_A/(n_A+n_B)$	$n_B/(n_A+n_B)$	0.0	0.0
single	0.5	0.5	0.0	-0.5
complete	0.5	0.5	0.0	0.5
weighted average	0.5	0.5	0.0	0.0
centroid	$n_A/(n_A+n_B)$	$n_B/(n_A+n_B)$	$-n_A n_B/(n_A+n_B)^2$	0.0
median	0.5	0.5	-0.25	0.0
Ward	$(n_A+n_i)/n_{ABi}$	$(n_B+n_i)/n_{ABi}$	$-n_i/n_{ABi}$	0.0

with $n_{ABi} = n_A+n_B+n_i$.

Non-hierarchical cluster analysis

With this method the object clusters are not hierarchically ordered but may be partioned independently of each other.

Commonly in nonhierarchical cluster analysis one starts with an initial partioning of objects to the different clusters. After that the membership of the objects to the clusters, e.g. to the cluster centroids, is determined and the objects are newly partioned. We consider here a general method for non-hierarchical clustering, that can be used for both crisp (classical) and fuzzy clustering, the *c-means algorithm*.

Initially beginning the objects are partioned into subsets S_i, where i is indexed from 1 to c – the number of clusters. The membership of an object with the feature vector x_k to cluster i can be characterized by means of a membership function, m_{ik}, as follows:

$$m_{ik} := m_{S_i}(x_k) \qquad (5\text{-}97)$$

For a crisp set the membership value is either 0 or 1. For fuzzy sets the membership of an object can assume values in the interval from 0 to 1 (cf. Sec. 8.3).

The matrix $M = [m_{ik}]$ is termed a *c-partition*, if the following conditions are fulfilled:

If methods of fuzzy clustering are used an object can belong to different clusters to a different extent. If crisp clustering is applied an object is uniquely assigned to a single cluster.

- The membership values of the n objects in the clusters are either crisp or fuzzy, i.e.

$$m_{ik} \in \{0,1\} \text{ or } [0,1] \quad 1 \le i \le c, 1 \le k \le n \qquad (5\text{-}98)$$

- The sum of the membership values of the objects in a given partition is equal to 1 for crisp sets. For fuzzy sets the sum is normalized to membership values of 1, i.e.

$$\sum_{i=1}^{c} m_{ik} = 1 \quad 1 \le k \le n \qquad (5\text{-}99)$$

- Within a given partition the objects must be partitioned among all the clusters, i.e. each cluster of a partition must contain at least one object. On the other hand in a 2-partition at most $n - 1$ objects can belong to a single cluster.

$$0 < \sum_{k=1}^{n} m_{ik} < n \qquad 1 \le i \le c \qquad \text{(5-100)}$$

Example 5-8: *Clustering (nonhierarchical)*

Consider as an example several 2-partitions for the three objects x_1, x_2, and x_3:

$$M_1 = \begin{bmatrix} 1 & 1 & 0 \\ 0 & 0 & 0 \end{bmatrix} \quad M_2 = \begin{bmatrix} 1 & 1 & 0 \\ 0 & 1 & 1 \end{bmatrix} \quad M_3 = \begin{bmatrix} 1 & 0 & 0 \\ 0 & 1 & 1 \end{bmatrix}$$

$$M_4 = \begin{bmatrix} 1 & 1 & 1 \\ 0 & 0 & 0 \end{bmatrix} \quad M_5 = \begin{bmatrix} 1 & 0 & 1 \\ 0 & 1 & 0 \end{bmatrix}$$

Each row represents a definite cluster and the columns characterize the objects, as they are assigned to the clusters. In this example there are only two genuine 2-partitions, i.e. M_3 and M_5. In the partition M_1 the object x_3 is not partitioned. In M_2 the object x_2 is twofold present, and in M_1 and M_4 the second cluster does not contain an object.

A *partition* is a definite assignment of objects to a given cluster.

As an example of fuzzy *c*-partitions the following partitions are given for the three objects:

$$M_1 = \begin{bmatrix} 1 & 0.5 & 0 \\ 0 & 0.5 & 1 \end{bmatrix} \quad M_2 = \begin{bmatrix} 0.8 & 0.5 & 0.2 \\ 0.2 & 0.5 & 0.8 \end{bmatrix}$$

$$M_4 = \begin{bmatrix} 0.8 & 0.9 & 0.3 \\ 0.2 & 0.2 & 0.7 \end{bmatrix} \quad M_5 = \begin{bmatrix} 0.8 & 1 & 0.9 \\ 0.2 & 0 & 0.1 \end{bmatrix}$$

As can be easily seen, in one of the four cases, i.e. M_3, the partition is incorrect – the sum of membership values exceeds 1 for object x_2.

To find genuine partitions the following scheme is applied:

- Characterization of the clusters by their centroids:

$$v_i = \frac{1}{\sum\limits_{k=1}^{n} m_{ik}} \sum_{k=1}^{n} m_{ik}^q x_k \qquad \text{(5-101)}$$

where v_i is the centroid of cluster i; q is an exponent expressing the degree of fuzziness, i.e. for $q = 1$ the classical c-means algorithm is obtained.

The results of a c-means clustering can be visualized by plotting the object-cluster distances. For the data in Table 5-4 we obtain:

- Computation of the differences between the objects and the cluster centroids by use of:

$$\|x_k - v_i\|^2 = \left[\sum_{j=1}^{p}(x_{kj} - v_i)^2\right]^{1/2} \tag{5-102}$$

where p is the number of variables.

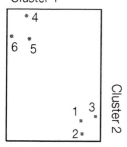

Cluster 1

- Minimization of the distance function by use of:

$$\min z(M,V) = \sum_{i=1}^{c}\sum_{k=1}^{n} m_{ik}\|x_k - v_i\|^2 \tag{5-103}$$

Minimization of the function $z(M, V)$ might become a computational problem. The number of all partitions to be tested is calculated according to:

$$\#\text{Partitions} = \frac{1}{c!}\left[\sum_{j=1}^{c}\binom{c}{j}(-1)^{c-j} j^n\right] \tag{5-104}$$

For e.g., for 10 clusters with 25 objects in total 10^{18} different partitions would have to be tested. Fortunately, not all partitions need to be computed, since algorithms are available that find an optimum partition iteratively according to predefined criteria.

Frequently a threshold is defined, in order to judge the improvement by changing from a partition M^l to M^{l+1}. If the criterion falls below the threshold the computation can be stopped.

5.2.3 Graphical methods

All the methods discussed so far for grouping of data cannot surpass the abilities of humans to recognize patterns. For this reason increasing number of methods are being developed that exploit the human ability for *recognizing patterns*. These graphical methods are based on the compact representation of multidimensionally characterized objects. We learn here about representation of multivariate data by means of star and sunray plots as well as by Chernoff faces.

For all those methods standardization and translation to positive feature values is a prerequisite.

Star and sun-ray plots

If the feature values are converted into polygons and the polar coordinates are plotted, then star-like representations emerge. Consider p features where a circle is segmented into p uniform sectors. Each sector contains an angle of $360°/p$. Each boundary line is assigned a feature. The actual value of the feature is then plotted at a distance from the mean point of the circle. After connection of the points by straight lines a polygon is obtained, that forms a characteristic pattern.

Graphical representation takes the form either of *stars* (Fig. 5-14, top), also termed diamonds, or the result is represented by drawing the boundaries as rays, giving the so-called sun-ray plots (Fig. 5-14, middle).

Feature representation
1

5

2

4

3

Chernoff faces

Assignment of features to facial features leads to respresentation of the objects as faces. The faces introduced by Chernoff are well known. The features are characterized by parameters such as the size or curvature of the eyes, the mouth, the eyebrows, the nose or the upper and lower halves of the face. As an example the hair data of the three subject groups are plotted in Fig. 5-14, bottom, as Chernoff faces.

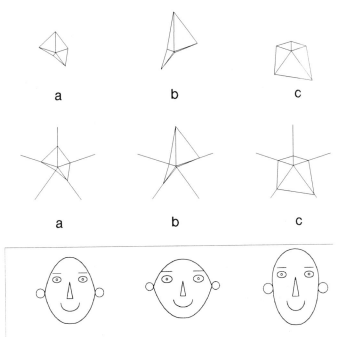

a   b   c

a   b   c

Fig. 5-14.
Graphical methods for grouping of the hair data from Table 5-1 for representatives of the three subject groups a, b and c (for assignments cf. Table 5-8). Top : stars; middle: sun-ray; bottom: Chernoff faces

Frequently up to 20 different facial parameters are used, as it is demonstrated in Fig. 5-15 for distinguishing between healthy and ill patients on the basis of clinical analyses of blood serum samples.

One disadvantage of Chernoff faces is that the individual faces cannot be varied independently of each other. Faces proposed by Flury-Riedwyl [cf. 5-1] overcome this disadvantage.

Healthy   Diseased

Fig. 5-15.
Chernoff faces for distinguishing between serum samples from diseased and healthy patients on the basis of 20 clinical analyses

5.3 Supervised methods

If the particular clusters of which objects are members are known in advance the methods of supervised pattern recognition can be used. Here the following methods are explained: linear learning machine, discriminant analysis, *k*-nearest neighbor, and the SIMCA method.

5.3.1 Linear learning machine

The first analytical application of a pattern recognition method dates back to 1969 when classification of mass spectra in respect of certain molecular-mass classes was tried with the *linear learning machine* (LLM). The basis of classification with the LLM is a discriminant function that divides the *n*-dimensional space into category regions that can be further used to predict the category membership of a test sample.

Consider the data in Table 5-6, the iodine content of hair samples from five different patients belonging to two categories. Since only one feature has been measured ($p = 1$) the data can be represented in 1-dimensional space as given in Fig. 5-16.

Table 5-6.
Iodine content of hair samples from different patients

Hair sample	Iodine content, ppm	Augmented component
1	0.29	1.0
2	4.88	1.0
3	0.31	1.0
4	3.49	1.0
5	4.46	1.0

To find a decision boundary that separates the two groups the data vectors must be augmented by adding a $(n + 1)$ component equal to 1.0. This ensures that the boundary separating the classes passes through the origin. If more than two categories must be separated several linear discriminant functions would have to be constructed.

Fig. 5-17.
LLM: representation of iodine data of Table 5-6 augmented by an additional dimension and separated by a straight-line boundary with the normal weight vector w

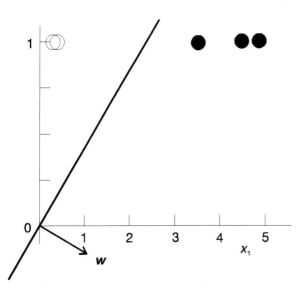

The boundary that separates the two categories is found iteratively by adjusting the elements of a weight vector, w, which is normal to the boundary, such that the dot product of w and any vector of the full circles is positive, while that of w and the empty circles is negative (Fig. 5-17). The decision boundary s is expressed by:

$$s = w_1 x_1 \tag{5-105}$$

or in general

$$s = w^T x = \|w\| \times \|x\| \cos \theta \tag{5-106}$$

where $\|.\|$ – vector norm, i.e. $\left[\sum_i x_i^2 \right]^{1/2}$

x – augmented data vector
s – scalar variable
θ – angle between w and x

161

If the angle θ is less than 90° it is obvious that the objects represented as full circles are categorized and that $s > 0.0$. Conversely if the angle θ is greater than 90° the scalar variable will be $s < 0$ and the empty circle objects are described.

To find the weight elements, w, they are set initially to random numbers. The objects are then classified by computing s and checking against the correct answer. If all classifications are correct the training process can be stopped and the LLM can be used for further classification purposes. However, if the response is incorrect new weights have to be calculated by updating the old ones, e.g. by use of:

$$w^{(new)} = w^{(old)} + cx \quad \text{with} \quad c = \frac{-2s}{x^T x} \tag{5-107}$$

Here the constant c is chosen such that the boundary is reflected to the correct side of a data point by the same distance as it was in error. The update of weights is repeated as long as all objects are correctly classified.

Of course, the LLM will work properly only if the data are linearly separable. One should also remember that the solution for positioning the boundary is not unique so that different solutions will emerge if the order of presentation of the training objects is changed.

Additional disadvantages are the simple class boundaries, the danger of wrong assignments of outliers, or slow convergence. In addition LLM is restricted to the separation of only two classes (binary classifier).

5.3.2 Discriminant analysis

LDA – Linear discriminant analysis

A more formal way of finding a decision boundary between different classes is based on LDA as introduced by Fisher and Mahalanobis. The boundary or hyperplane is calculated such that the variance between the classes is maximized and the variance within the individual classes is minimized. There are several ways of arriving at the decision hyperplanes. In fact, one of the routes Fisher described can be understood from the principles of straight line regression (cf. Sec. 6.1).

A decision boundary separates two or more groups of data.

Two-class case

Consider again the iodine data in Table 5-6. If you create an augmented variable y with values +1 for objects from one group and −1 for objects from the other group and perform a linear regression (LR) of y on x for the five training objects then the regression equations for y will have coefficients similar to the LDA weights, i.e.,

$$y = -1.082 + 0.477\,x$$

Comparison with the LLM reveals that the boundary is not constrained to pass through the origin but a cut-off point is included into the model. Application of LDA gives the discriminant function

$$s = -4.614 + 1.718\,x$$

Both boundaries are given in Fig. 5-18.

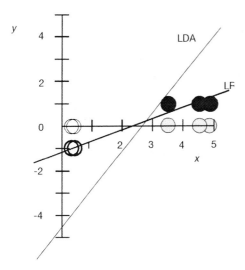

Fig. 5-18.
Decision lines based on LR and LDA for separating the objects represented by full and empty circles

Multi-class case

To arrive at the (nonelemental) LDA solution an eigenvalue problem must be solved. To generalize the problem we consider a data matrix X with n objects and p feature variables. There are g different groups or classes indexed by g_1 to g_{n_j} :

$$\begin{bmatrix}
x_{11} & x_{12} & \cdots & x_{1p} \\
x_{21} & x_{22} & & x_{2p} \\
x_{31} & x_{32} & \cdots & x_{3p} \\
x_{41} & x_{42} & & x_{4p} \\
\vdots & & & \\
x_{j1} & x_{j2} & \cdots & x_{jp} \\
x_{n1} & x_{n2} & & x_{np}
\end{bmatrix}
\begin{matrix}
\left.\vphantom{\begin{matrix}x\\x\end{matrix}}\right\}g_1 \\
\left.\vphantom{\begin{matrix}x\\x\end{matrix}}\right\}g_2 \\
\\
\left.\vphantom{\begin{matrix}x\\x\end{matrix}}\right\}g_{n_j}
\end{matrix} \qquad (5\text{-}108)$$

163

The weights of the linear discriminant functions are found as the eigenvectors of the following matrix:

$$G^{-1}Hw = \lambda w \qquad (5\text{-}109)$$

where λ – eigenvalue

The results of categorial classification are represented in the *confusion matrix*, that contains the numbers of correctly classified objects in each class on the main diagonal and the misclassified objects in the off-diagonal.

The matrix G is derived from the covariance matrix C (Eq. 5-11) of the different classes or groups g as follows:

$$G = (n-g)\,C = (n-g)\,\frac{1}{n-g}\sum_{j=1}^{g}(n_j - 1)\,C_j \qquad (5\text{-}110)$$

$$C_j = \frac{1}{n_j - 1}\sum_{l\in g_j}(x_{li} - \bar{x}_{ji})(x_{lk} - \bar{x}_{jk}) \qquad (5\text{-}111)$$

where

n – total number of objects
n_j – number of objects in group j
$l\in g_j$ – index l is an element of the jth group g_j

Matrix H describes the spread of the group means x_j over the grand average $\bar{x}$, i.e.

$$H = \sum_{j=1}^{g} n_j\,(\bar{x}_j - \bar{x})(\bar{x}_j - \bar{x})^{\mathrm{T}} \qquad (5\text{-}112)$$

$$\bar{x} = \frac{1}{n}\sum_{j=1}^{n} n_j \bar{x}_j \qquad (5\text{-}113)$$

The eigenvector, w_1, found on the basis of the greatest eigenvalue λ_1 provides the first linear discriminant function, s_1, from:

$$s_1 = w_{11}x_1 + w_{12}x_2 + ... + w_{1p}x_p \qquad (5\text{-}114)$$

With the residual x-data the second largest eigenvalue is computed and with the new eigenvector, w_2, the second discriminant function, s_2, is obtained:

$$s_2 = w_{21}x_1 + w_{22}x_2 + ... + w_{2p}x_p \qquad (5\text{-}115)$$

The procedure is continued until all discriminant functions needed to solve the discrimination problem are found. By plotting pairs of discriminating functions against each other the best separation of objects into groups after linear transformation of the initial features can be visualized (cf. Fig. 5-19). The projection of a particular object on to the separating line or hyperplane is called its *score* on the linear discriminant function.

Classification of unknown objects, x_u, is carried out by inserting the feature data into the discriminant functions in order to transform its coordinates in the same way as for the original data set. Then, the object is assigned to that class for which its centroid has the smallest Euclidian distance:

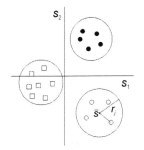

$$\min_{j} \left\| w^T (x_u - \bar{x}_j) \right\| \quad \text{with } j = 1, ..., g \tag{5-116}$$

Example 5-9: *Linear discriminant analysis*

LDA is to be used to build a classification model for the hair data given in Table 5-1. On the basis of the model an unknown hair sample as given in Table 5-7 is to be classified.

Table 5-7.
Elemental content, in ppm, of an unknown hair sample

Cu	Mn	Cl	Br	I
9.2	0.27	2200	9.8	4.7

Table 5-8.
Classification vector
for the hair samples in Table 5-1

Hair No.	Subject
1	B
2	A
3	C
4	B
5	A
6	C
7	C
8	A
9	B

In the first step the discriminant function is computed by use of Eqs. (5-109) to (5-113). The allocation of hair samples to the individual groups is represented by a classification vector as given in Table 5-8.

For the first two discriminant functions we get:

$$s_1 = 0.227 x_{Cu} + 0.694 x_{Mn} - 1.200 x_{Cl} + 0.0394 x_{Br} - 0.0514 x_I \tag{5-117}$$

$$s_2 = 0.00672 x_{Cu} + 0.936 x_{Mn} - 0.211 x_{Cl} + 1.342 x_{Br} - 0.395 x_I \tag{5-118}$$

The first discriminant function describes 63.39% of the variance of the data and the second one 36.61%. That is, the two discriminant functions are sufficient to explain 100% (63.39% + 36.61%) of the data variance. Discrimination of the data into three classes can be illustrated by plotting the discriminant functions against each other (Fig. 5-19).

If in the second step the elemental content of the unknown hair sample (Table 5-7) is inserted into the discriminant functions, Eqs. 5-117 and 5-118, the following values then result for the scalar values: $s_1 = 4.416$ and $s_2 = -7.15$. By comparison with the centroids of the classes (Fig. 5-19) we obtain for the Euclidian distances in relation to the three classes the values 15.66, 4.93, and 10.52 for the classes A, B, and C, respectively. The shortest distance is found between the data of the unknown hair and class B, i.e. the hair must be assigned to subject group B.

165

Fig. 5-19.
Discriminant function on the basis
of linear discriminant analysis
of the hair data in Table 5-1 on the
basis of the classification vector
in Table 5-8. The centroids
of the classes are labeled '+'.

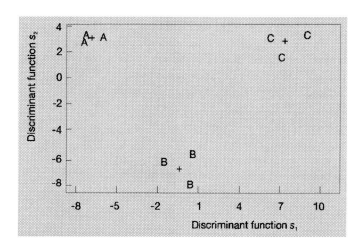

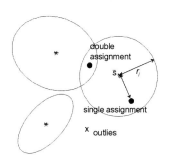

Multiple-class assignment

Two disadvantages of this method of unique decision are that simultaneous membership of an object to several classes is not detected and that outliers, which do not belong to any of the classes, will always be categorized. Therefore, the unique categorization is often replaced by assigning the object to all classes within a fixed variance range, e.g. 95%. If the object lies outside any of those variance ranges it will not be categorized.

The variance radius, r_j, is calculated from:

$$r_j = \frac{d(n-g)}{n-g-d+1} \frac{n_j+1}{n_j} F_{(d,n-g-d+1;\alpha)} \qquad (5\text{-}119)$$

where d – number of used discriminant functions
          F – Fishers F-statistic with risk α

An object is assigned to a particular class if:

$$\sum_{i=1}^{d}(s_i - \bar{s}_{ij})^2 \le r_j \qquad (5\text{-}120)$$

Here s_i represents the new coordinates of the unknown object and $\bar{s}_{ij}$ is the jth class centroid.

Necessary *assumptions* of LDA are the normality of data distributions, the existence of different class centroids, and the similarity of variances and covariances among the different groups. Therefore, classification problems arise if the variances of groups differ substantially or if the direction of objects in the pattern space is different, as depicted in Fig. 5-20.

166

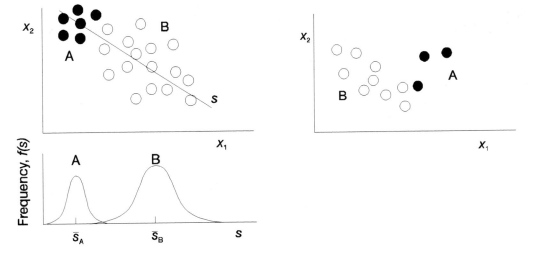

Fig. 5-20. Differently spread (top left) and differently directed (right) objects of the full and empty circled classes. For the left case the density function around the class centroids is given.

Bayesian classification

If the objects of all classes obey a multivariate normal distribution, an optimal classification rule can be based on *Bayes's theorem*. The assignment of a sample, x, characterized by p features to a class j of all classes g is based on maximizing the *posterior probability*

$$P(j \mid x) \quad \text{for} \quad k = 1, ..., g \tag{5-121}$$

Application of Bayes's theorem for calculation of the posterior probability reveals:

$$P(j \mid x) = \frac{p(x \mid j) P(j)}{p(x)} \tag{5-122}$$

According to Eq. 5-122 the posterior probability is computed from the probability density function for the considered class, $p(x \mid j)$, the prior probability for that class $P(j)$ and the probability density function over all classes $p(x)$. A sample x is then assigned to that class j, for which the largest posterior probability is found.

For computation of the class probability density, $p(x \mid j)$, the multidimensional normal distribution is assumed:

$$p(x \mid j) = (2\pi)^{-d/2} \left| S_j \right|^{-0.5}$$
$$\times \exp\left[-0.5(x - \bar{x}_j) S_j^{-1} (x - \bar{x}_j)^{\mathrm{T}} \right] \tag{5-123}$$

167

where the covariance matrix S_j based on the class centroid $\bar{x}_j$ is obtained from:

$$S_j = \frac{1}{n_j} \sum_{i=1}^{n_j} (x_i - \bar{x}_j)^{\mathrm{T}} (x_i - \bar{x}_j) \tag{5-124}$$

$$\bar{x}_j = \frac{1}{n_j} \sum_{i=1}^{n_j} x_i^{(j)} \tag{5-125}$$

n_j describes the number of samples in class j.

Maximizing the posterior probability is related to minimizing the discriminant scores obtained from:

$$d_j(x) = (x - \bar{x}_j) S_j^{-1} (x - \bar{x}_j)^{\mathrm{T}} + \ln|S_j| - 2\ln P(j) \tag{5-126}$$

An unknown sample is assigned to the class j, for which the distance to its class centroid is shortest. The first term in Eq. 5-126 represents the Mahalanobis distance between the sample x and the class centroid $\bar{x}_j$.

In LDA it is assumed that the class covariance matrices are equal, i.e. $S_j = S$ for all $j = 1$ to g. Different class covariances are allowed in quadratic discriminant analysis (QDA). The results are quadratic class boundaries based on unbiased estimates of the covariance matrix. The most powerful method is based on regularized discriminant analysis (RDA) [5-7]. This method seeks biased estimates of the covariance matrices, S_j, to reduce their variances. This is done by introducing two regularization parameters λ and γ according to:

$$S_j(\lambda) = (1 - \lambda) S_j + \lambda S \tag{5-127}$$

$$S_j(\lambda, \gamma) = (1 - \gamma) S_j(\lambda) + \gamma\, tr[S_j(\lambda)]/pS \tag{5-128}$$

RDA corresponds to QDA, if $\lambda = 0$ and $\gamma = 0$. For $\lambda = 1$ and $\gamma = 0$ the method corresponds to LDA. If $\lambda = 1$ and $\gamma = 1$, RDA is the same as the *nearest mean classifier*.

The parameters range in the interval 0 and 1; tr characterizes the trace of a matrix.

This ensures that even for ill-conditioned systems, e.g. very similar spectra, good results are to be expected for the estimation of the inverse covariance matrix in Eq. 5-126 and the subsequent classification of unknown samples.

5.3.3 *k*-nearest neighbor method

The *k*-NN method is also used for filling in missing values or in library searches.

The *k*-nearest neighbor method introduced by Fix and Hodges in 1951 is a simple nonparametric classification method. For classification of an unknown object, its distance, usually the

Euclidian distance, from all objects is computed. The minimum distance is selected and the object is assigned to the corresponding class.

Typically the number of neighboring objects k is chosen to be 1 or 3. With the k-NN method, very flexible separation boundaries are obtained as exemplified in Fig. 5-21.

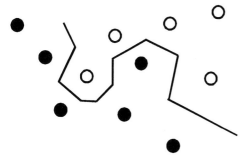

Fig. 5-21.
Separation boundary
for classification of objects
into two classes with $k = 1$

Unfortunately, classification is dependent on the number of objects in each class. When classes overlap the object will be assigned to the class with the larger number of objects. This situation can sometimes be handled if no single criterion is used but one allows alternative counting of neighborhood, e.g., for a class A with fewer objects five neighbors must be found, whereas for another class B with more objects seven neighbors would have to be considered.

5.3.4 SIMCA

Apart from discrimination methods, class membership of objects can also be determined by description of the individual classes by means of a separate mathematical model independent of the model for the other classes. In terms of geometry the model describes an envelope or a 'class box' around the class so that unknown objects can be classified according to their fit to a particular class model.

An early developed method uses multivariate normal distribution to model the classes on the basis of their data variances. Although this model has sometimes been used in analytical chemistry it lacks general applicability because the method is based on the covariance matrix where it is tacitly assumed that many data exist and that the ratio between objects and variables is favorable – approximately 6 to 1.

More often the SIMCA method is used which finds separate principal component models for each class. By using SIMCA, the object variable number ratio is less critical and the model is constructed around the projected rather than the original data. The basic steps of principal component calculations as needed

SIMCA is the abbreviation for soft *independent modeling of class analogies*.

for SIMCA have been outlined in the chapter on projection methods with the NIPALS algorithm (Sec. 5.1.2).

For each class q a separate model is constructed that reveals for a single x-observation:

$$x_{ij}^q = \bar{x}_j^q + \sum_{a=1}^{A_q} t_{ia}^q l_{ja}^q + e_{ij}^q \tag{5-129}$$

where

x_{qj} – mean of variable j in class q

A_q – number of significant principal components in class q

t_{ia}^q – score of object i on component a in class q

p_{ja}^q – loading of variable j on principal component a in class q

e_{ij}^q – residual error of object i and variable j

Fig. 5-22.
SIMCA models for different numbers of significant principal components

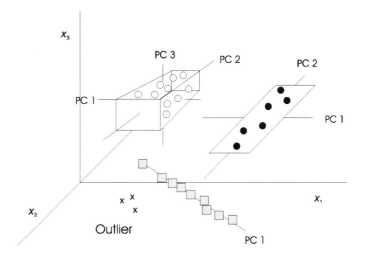

The number of principal components in a class model is determined by *cross validation*.

The principal components are found by means of the iterative NIPALS algorithm. Each separate model might reveal a different number of significant principal components A_q. Thus, the class models might represent lines, planes, boxes or hyperboxes as demonstrated in Fig. 5-22. The models can then be used to eliminate outlying objects, to estimate the modeling power of a particular variable, and to classify new objects.

Outliers

Objects that do not fit the estimated principal component model can be eliminated by testing the total residual variance of a class q against the residual variance of that object. The two variances are calculated as follows:

Total residual variance of class q

$$s_0^2 = \sum_{i=1}^{n} \sum_{j=1}^{p} \frac{e_{ij}^2}{(n - A_q - 1)(p - A_q)} \qquad (5\text{-}130)$$

where n is the number of objects, and p is the number of variables.

Residual variance for object i

$$s_i^2 = \sum_{j=1}^{p} \frac{e_{ij}^2}{p - A_q} \qquad (5\text{-}131)$$

If both variances are of the same order of magnitude the object is assigned as a typical object to the class. If $s_i^2 > s_0^2$ then the object should be eliminated to enable a more parsimonious class model to be described.

Modeling power

The residual variance of a variable j of class q is used for estimating the modeling power of a particular variable if related to the so-called meaningful variance – the familiar expression for the variance:

Residual variance of variable j

$$s_j^2 (\text{error}) = \sum_{i=1}^{n} \frac{e_{ij}^2}{n - A_q - 1} \qquad (5\text{-}132)$$

Meaningful variance of variable j

$$s_j^2 (x) = \sum_{i=1}^{n} \frac{(x_{ij} - \bar{x})^2}{n - 1} \qquad (5\text{-}133)$$

Comparison of the two variances reveals a measure for the noise-to-signal ratio of this variable. The modeling power for variable j, R_j, is derived from the following expression:

$$R_j = 1 - \frac{s_j (\text{error})}{s_j (x)} \qquad (5\text{-}134)$$

If the modeling power approaches values of 1 then the variable will be highly relevant, because the ratio between the residual error for the variable is small compared with its meaningful variance.

Classification

An unknown object with data vector x_u (dimension $1 \times p$) is checked for membership of a particular class by regression of the vector x_u on the q class models. Multiplying the data vector by the loading matrix L $(p \times A_q)$ gives an estimate of a new score vector $\hat{t}$ $(1 \times p)$. With the score vector the residuals are computed and used to decide on the membership of the object to a class:

$$\hat{t} = x_u L \qquad (5\text{-}135)$$

$$e = x_u - \hat{t} L^T \qquad (5\text{-}136)$$

As the residual variance for object u we get:

$$s_u^2 = \sum_{j=1}^{p} \frac{e_{uj}^2}{p - A_q} \qquad (5\text{-}137)$$

The object is assigned to class q if the variances s_u^2 and s_0^2 are of similar orders of magnitude. If s_u^2 is greater than s_0^2 the object is not a member of class q.

5.4 General reading

[5-1] Varmuza, K., *Pattern recognition in chemistry*, Berlin: Springer, 1980.
[5-2] Malinowski, E. R., Howery, D. G., *Factor analysis in chemistry*, New York: Wiley, 1980.
[5-3] Frank, I. E., Todeschini, R., *The data analysis handbook*, Amsterdam: Elsevier, 1994.
[5-4] Henrion, R., Henrion, G., *Multivariate data analysis*. Berlin: Springer, 1994.
[5-5] Massart, D. L., Vandeginste, B. G. M., Deming, S. N., Michotte, Y., L. Kaufmann, *Chemometrics – a textbook*, Amsterdam: Elsevier, 1988.
[5-6] Sharaf, M. A., Illman, D. L., Kowalski, B. R., *Chemometrics*. Chemical Analysis Series Vol. 82, New York: Wiley, 1986.
[5-7] Friedman, J. H., *Regularized Discriminant Analysis*, J. Am. Stat. Assoc. **84** (1989) 165.

Questions and problems

1. Explain the following methods for data preprocessing: centering, range scaling, autoscaling, scaling to variance one, normalization, FT-transformation, principal component projection, linear transformation, logarithmic transformation.

2. Specify the following terms in multivariate analysis: principal component, eigenvector, common and unique factor, score, loading, target vector, latent variable.
3. Which methods can be used to decide on the number of abstract components in principal component analysis?
4. What is the difference between principal component and factor analysis?
5. The heavy metal content of soil samples has been determined at nine different locations. The table below provides the results of the analysis for the elements Zn, Cd, Pb, and Cu (ppm):

Sample	Zn	Cd	Pb	Cu
1	35.3	0.08	0.25	6.5
2	20.2	1.20	0.52	3.2
3	34.2	0.05	0.28	5.8
4	22.2	1.50	0.48	2.9
5	33.8	0.07	0.26	4.9
6	25.3	0.90	0.60	3.6
7	38.1	2.10	1.20	3.0
8	39.2	1.90	1.50	2.5
9	37.8	2.80	1.40	2.6

(a) What are the eigenvalues of the data matrix?
(b) How many significant principal components/factors are in the data set if more than 98 % of the total variance is explained?
(c) Which sample numbers belong to the classes found?
(d) Assign the following unknown sample to the appropriate class of origin: Zn 21.8 ppm; Cd 1.3 ppm; Pb 0.5 ppm; Cu 3.3 ppm.
6. Summarize the advantages and disadvantages of linear, quadratic and regularized discriminant analysis compared to the SIMCA method.

6 Modeling

Learning objectives

■ To introduce univariate regression analysis for straight-line calibration and empirical model-building of analytical relationships
■ To understand the usage of analysis of variance and of regression diagnostics
■ To model analytical relationships by multiple linear regression analysis, such as in multicomponent analysis, and by target transform factor analysis
■ To present nonlinear regression methods based on parametric and nonparametric models

Models are constructed in analytical chemistry to describe the *relationship* between *reponses* and *factors*. This is, e.g., important for the optimization of analytical methods on the basis of **R**esponse **S**urface **M**ethods (cf. Sec. 4.2.3). On the other hand, models are needed for calibration of analytical methods. There calibration of the dependence of the amount of a single analyte on one or several wavelengths might be of interest. If, in the first example, the straight-line model is adequate, for the second task of multi-wavelength spectroscopy multivariate approaches are needed. On the other hand, calibrations must be performed for unselective analytical methods. These methods are termed simultaneous *multi-component analysis*. In NIR spectroscopy the water and protein content of whole-grain wheat are determined that way.

Calibration and *Response Surface Methods* are indeed the most important applications of regression methods in analytical chemistry. Other applications are seen in environmental analysis, where receptor models are developed on the basis of multivariate relationships or, for risk assessment of environmental pollution, on the basis of typical pollution patterns.

Although our world in general cannot be described by simple linear relationships, *linear models* are applicable to most of the problems considered in the following. More precisely, models, that are linear in the parameters to be estimated are sufficient. The simplest linear model is the straight-line model. For one dependent variable, y, and one independent variable, x, with the intercept b_0 and the slope b_1 the model is expressed by:

$$y = b_0 + b_1 x \qquad (6\text{-}1)$$

Linear models consist of additive terms, each of which contains only a multiplicative parameter.

However, the quadratic dependence of one variable y on another variable x can also be characterized by a linear model, i.e.:

$$y = b_0 + b_1 x + b_{11} x^2 \tag{6-2}$$

All parameters (b_0, b_1 and b_{11}) can be estimated by the methods of linear algebra. Since the calculation of the parameters in multivariate linear models can be carried out by the same principles as are used for the straight-line model, we will start our considerations with the problem of univariate linear regression.

6.1 Univariate linear regression

Straight-line model

The straight-line model (6-1) has already been used as a calibration function in Sec. 4.1 (cf. Fig. 4-2 and Eq. 4-1). Estimation of the parameters b_0 and b_1 by means of linear regression for n measurements of the pairs of values (x_i, y_i) is achieved as follows:

$$b_1 = \frac{n \sum_{i=1}^{n} x_i y_i - \sum_{i=1}^{n} x_i \sum_{i=1}^{n} y_i}{n \sum_{i=1}^{n} x_i^2 - \left(\sum_{i=1}^{n} x_i \right)^2} \tag{6-3}$$

$$b_0 = \bar{y} - b_1 \bar{x} \tag{6-4}$$

$$\text{where} \quad \bar{x} = \frac{1}{n} \sum_{i=1}^{n} x_i \tag{6-5}$$

$$\bar{y} = \frac{1}{n} \sum_{i=1}^{n} y_i \tag{6-6}$$

The variances of the parameters are estimated according to:

$$s_{b_0}^2 = \frac{s_y^2 \sum_{i=1}^{n} x_i^2}{n \sum_{i=1}^{n} (x_i - \bar{x})^2} \tag{6-7}$$

$$s_{b_1}^2 = \frac{s_y^2}{\displaystyle\sum_{i=1}^{n}(x_i - \bar{x})^2} \tag{6-8}$$

s_y^2 characterizes the mean variance of *residuals* as differences between the measured values, y, and the values predicted by the model (Eq. 6-1), $\hat{y}_i$. The residual variance is calculated from:

$$s_y^2 = \frac{\displaystyle\sum_{i=1}^{n}(y_i - \hat{y}_i)^2}{n-p} \qquad \text{with} \qquad p = 2 \tag{6-9}$$

The parameter estimates are frequently given as the standard error, i.e. as s_{b_0} or s_{b_1} (cf. Eq. 6-23).

For *prediction* of an x_0-value from a y_0-value by means of the straight-line model we have:

$$x_0 = \frac{y_0 - b_0}{b_1} \tag{6-10}$$

The standard deviation for predictions obtained by use of the straight-line model and performing p parallel measurements with one sample gives:

$$s_0 = \frac{s_y}{b_1}\sqrt{\frac{1}{p} + \frac{1}{n} + \frac{(\bar{y}_0 - \bar{y})^2}{b_1^2 \displaystyle\sum_{i=1}^{n}(x_i - \bar{x})^2}} \tag{6-11}$$

It is important for all kind of modeling that the estimated parameters are tested for their statistical significance. To derive appropriate tests for the adequacy of a regression model we need to generalize the straight-line regression.

Generalization of the straight-line model

For generalization of the regression problem we use matrix notation. The straight-line model

$$y = b_0 \cdot 1 + b_1 x \tag{6-1}$$

then reads:

$$\begin{pmatrix} y_1 \\ y_2 \\ \vdots \\ y_n \end{pmatrix} = \begin{pmatrix} 1 & x_1 \\ 1 & x_2 \\ \vdots & \vdots \\ 1 & x_n \end{pmatrix} \begin{pmatrix} b_0 \\ b_1 \end{pmatrix} \tag{6-12}$$

or in abbreviated notation:

$$y = Xb \qquad (6\text{-}13)$$

The number of rows of the vector of the dependent variable y and of the matrix of the independent variables X corresponds to the number of measurements n. The parameter vector b consists in the present case of the straight line equation of only two elements.

The parameter elements in vector b are estimated on the basis of the generalized inverse from:

$$b = (X^T X)^{-1} X^T y \qquad (6\text{-}14)$$

The most appropriate methods for solving Eq. 6-14 will be discussed in Sec. 6.2.1. The statistical tests for the adequacy of the regression model are based on analysis of variances (cf. Sec. 2.3).

ANOVA – Analysis of variance

For ANOVA the observed and predicted y-values are considered in terms of their dependence on the independent variables or factors (cf. Sec. 2.3).

Partitioning of the variances of a linear regression follows the scheme given in Table 6-1 and Fig. 6-1 by the appropriate sums of squares: the total variance of the y-values, SS_T, is derived from the sum of squares of the mean, SS_M, and the sum of squares corrected for the mean, SS_{corr}.

The latter sum of squares constitutes from the sum of squares of the factors, SS_{fact}, and the sum of squares of the residuals, SS_R. The sum of squares of the residuals is composed of the sum of squares of the lack of fit, SS_{lof}, and the sum of squares of pure experimental error, SS_{pe}.

Table 6-1.

Computation of the sums of squares (*SS*) for a complete ANOVA in linear regression

Sum of Squares	Matrix operation	Calculation	Degrees of freedom
SS_T, Total	$y^T y$	$\sum_{i=1}^{n} y_i^2$	n
SS_M, Mean	$\bar{y}^T \bar{y}$	$n\bar{y}^2$	1
SS_{corr}, Corrected for the mean	$(y-\bar{y})^T (y-\bar{y})$	$\sum_{i=1}^{n}(y_i - \bar{y})^2$	$n-1$
SS_{fact}, Factors	$(\hat{y}-\bar{y})^T (\hat{y}-\bar{y})$	$\sum_{i=1}^{n}(\hat{y}_i - \bar{y})^2$	$p-1$
SS_R, Residuals	$(y-\hat{y})^T (y-\hat{y})$	$\sum_{i=1}^{n}(y_i - \hat{y}_i)^2$	$n-p$
SS_{lof}, Lack-of-fit	$(j-\hat{y})^T (j-\hat{y})$	$\sum_{i=1}^{n}(\bar{y}_i - \hat{y}_i)^2$	$f-p$
SS_{pe}, Pure experimental error	$(y-j)^T (y-j)$	$\sum_{i=1}^{n}(y_i - \bar{y}_i)^2$	$n-f$

j – This vector contains *y*-values averaged at each observation point i. f – number of different realizations of the independent variables (independent factor combinations); n – number of measurements; p – number of parameters; $\bar{y} = \frac{1}{n}\sum_{i=1}^{n} y_i$ – total mean

In Example 6-1 we consider the calculation of all these sums of squares for a simple regression problem. The meaning of the different *y*-expressions is illustrated in Fig. 6-2.

Example 6-1: *ANOVA in regression analysis*

In Table 6-2 the *x-y*-data are given for four measurements. The data must be fitted by the straight-line model. After that all sums of squares are to be calculated according to Table 6-1, in order to use them for subsequent tests of the adequacy of the model.

Table 6-2.
x-y-data

No.	x	y
1	0	0.3
2	1	2.2
3	2	3
4	2	4

179

Insertion of the data in the model for the straight line Eq. 6-12 gives:

$$\begin{pmatrix} 0.3 \\ 2.2 \\ 3 \\ 4 \end{pmatrix} = \begin{pmatrix} 1 & 0 \\ 1 & 1 \\ 1 & 2 \\ 1 & 2 \end{pmatrix} \begin{pmatrix} b_0 \\ b_1 \end{pmatrix} \tag{6-15}$$

Computation of the parameter vector b is performed according to Eq. 6-14:

$$\begin{pmatrix} b_0 \\ b_1 \end{pmatrix} = b = (X^T X)^{-1} X^T y = \left(\begin{pmatrix} 1 & 1 & 1 & 1 \\ 0 & 1 & 2 & 2 \end{pmatrix} \begin{pmatrix} 1 & 0 \\ 1 & 1 \\ 1 & 2 \\ 1 & 2 \end{pmatrix} \right)^{-1} \begin{pmatrix} 1 & 1 & 1 & 1 \\ 0 & 1 & 2 & 2 \end{pmatrix} \begin{pmatrix} 0.3 \\ 2.2 \\ 3 \\ 4 \end{pmatrix}$$

$$= \begin{pmatrix} 4 & 5 \\ 5 & 9 \end{pmatrix}^{-1} \begin{pmatrix} 9.5 \\ 16.2 \end{pmatrix} = \begin{pmatrix} 0.818 & -0.454 \\ -0.454 & 0.363 \end{pmatrix} \begin{pmatrix} 9.5 \\ 16.2 \end{pmatrix} = \begin{pmatrix} 0.41 \\ 1.57 \end{pmatrix} \tag{6-16}$$

The straight line model is then:

$$y = 0.41 + 1.57x$$

On the basis of the model the sums of squares can be determined by ANOVA:

The variance of the y-values results from the *total sum of squares*, SS_T, i.e:

$$SS_T = y^T y = (0.3 \quad 2.2 \quad 3 \quad 4) \begin{pmatrix} 0.3 \\ 2.2 \\ 3 \\ 4 \end{pmatrix} = \sum_{i=1}^{n} y_i^2 = 0.3^2 + 2.2^2 + 3^2 + 4^2 = 29.93$$

The sum of squares of the *mean*, SS_M, is calculated from the total mean for all y-values by use of the equation:

$$SS_M = n\bar{y}^2 = 4 \cdot 2.375^2 = 22.56$$

Correspondingly we obtain for the sum of squares *corrected for the mean*:

$$SS_{corr} = \sum_{i=1}^{n} (y_i - \bar{y})^2 = (0.3 - 2.375)^2 + (2.2 - 2.375)^2 + (3 - 2.375)^2 + (4 - 2.375)^2 = 7.37$$

The influence of the *factors* is reflected by the sum of squares due to the *factors*, SS_{fact}:

$$SS_{fact} = \sum_{i=1}^{n} (\hat{y}_i - \bar{y})^2 = (0.41 - 2.375)^2 + (1.98 - 2.375)^2$$
$$+ (3.55 - 2.375)^2 + (3.55 - 2.375)^2 = 6.80$$

The difference between the observed and predicted y-values leads to the sum of squares of the *residuals*, SS_R (cf. Eq. 6-9):

$$SS_R = \sum_{i=1}^{n} (y_i - \hat{y}_i)^2 = (0.3 - 0.41)^2 + (2.2 - 1.98)^2$$
$$+ (3 - 3.55)^2 + (4 - 3.55)^2 = 0.57$$

Within the residuals the lack-of-fit to the model and the experimental error must be accounted for. The sum of squares due to lack-of-fit, SS_{lof}, is given by:

If the intercept is included in the model the *sum of residuals* equals zero.

$$SS_{lof} = \sum_{i=1}^{n} (\bar{y}_i - \hat{y}_i)^2 = (0.3 - 0.41)^2 + (2.2 - 1.98)^2$$
$$+ (3.5 - 3.55)^2 + (3.5 - 3.55)^2 = 0.07$$

The sum of squares due to *experimental error*, SS_{pe}, is calculated from:

$$SS_{pe} = \sum_{i=1}^{n} (y_i - \bar{y})^2 = (0.3 - 0.3)^2 + (2.2 - 2.2)^2$$
$$+ (3 - 3.5)^2 + (4 - 3.5)^2 = 0.5$$

Table 6-3 summarizes the results of the calculations in an ANOVA-table.

Table 6-3.
ANOVA-table for linear regression of the data in Table 6-2

Source of Variation	SS	df	MSS	F-value	p-value
Total, SS_T	29.93	4	7.48		
Mean, SS_M	22.56	1	22.56		
Corrected for the mean, SS_{corr}	7.37	3	2.47		
Factors, SS_{fact}	6.80	1	6.80	24.06	0.0391
Residuals, SS_R	0.57	2	0.283		
Lack-of-fit, SS_{lof}	0.07	1	0.07	0.14	0.779
Pure experimental error, SS_{pe}	0.50	1	0.5		

SS – sum of squares, MSS – mean sum of squares, df – degrees of freedom

Fig. 6-1.
Illustration of ANOVA for linear
regression analysis of the data
in Table 6-2

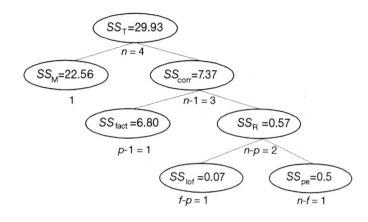

Fig. 6-2.
Plot of the *x-y*-data in Table 6-2
in conjunction with ANOVA
according to Table 6-1

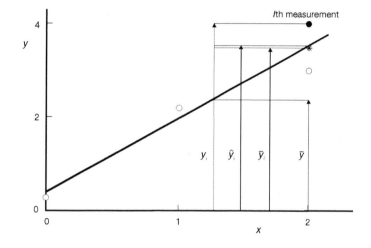

The *experimental error* can
only be estimated if at least one
replicate analysis is performed
at an independent factor
combination.

Coefficient of determination and correlation coefficient

The coefficient of determination is given by the ratio of the sums of squares due to the factors and to the sum of squares corrected for the mean:

If the regression parameters are
estimated by the least-squares
method, the square root of the
coefficient of determination and
the multiple *correlation coefficient*
will be identical.

$$R^2 = \frac{QS_{fact}}{QS_{corr}} \tag{6-17}$$

The coefficient of determination describes the fraction of the sum of squares due to the factors relative to the sum of squares corrected for the mean. The square root of the coefficient of determination gives the multiple correlation coefficient:

$$r = \sqrt{\frac{QS_{fact}}{QS_{corr}}} \tag{6-18}$$

The sign of the correlation coefficient is given by the slope b_1.

Example 6-2: *Correlation coefficient*

For the regression problem in Example 6-1 we obtain according to Eq. 6-17 for the coefficient of determination:

$$R^2 = \frac{SS_{fact}}{SS_{corr}} = \frac{6.80}{7.37} = 0.923$$

The two factors explain 92.3% of the sum of squares corrected for the mean.
The correlation coefficient is calculated from Eq. 6-18:

$$r = \sqrt{\frac{SS_{fact}}{SS_{corr}}} = \sqrt{\frac{6.80}{7.37}} = 0.961$$

Note that the correlation coefficient does not provide a realistic picture of the explanation of the variance due to the factors. In our example only 92.3% is explained, although the correlation coefficient is 0.961. An r-value of, e.g., 0.7 would mean that only 49% of the SS_{corr} can be explained.

Test for adequacy of the model

To test for the fit of data to the model two F-tests must be carried out: the F-test for goodness-of-fit and the F-test for lack-of-fit. For this the sum of squares normalized to the degrees of freedom is applied; this is given in Table 6-3 as the mean sum of squares (MSS).

As a measure of the quality of fit the F-value for the *goodness-of-fit* is calculated for the linear model with the intercept, b_0, from:

$$F(p-1, n-p) = \frac{SS_{fact}/(p-1)}{SS_R/(n-p)} = \frac{MSS_{fact}}{MSS_R} \qquad (6\text{-}19)$$

The mean sum of squares must be greater than the residuals if the factors in the model definitely influence the y-values. If an appropriate model has been found, the F-test will be *significant* at the given α-level.

The *lack-of-fit test* is based on a comparison of the mean sum of squares due to the model and the experimental error:

$$F(f-p, n-f) = \frac{SS_{lof}/(f-p)}{SS_{pe}/(n-f)} = \frac{MSS_{lof}}{MSS_{pe}} \qquad (6\text{-}20)$$

If the experimental error is comparably large for all factor combinations, the data are termed *homoskedastic*.

For *heteroskedastic data* the errors differ at different factor combinations.

183

In regression analysis the term *multiple realizations* of an x-value is common instead of factor combinations.

Note that this test is only applicable if the number of independent factor combinations, f, is greater than the number of model parameters, p. In addition at least one replicate measurement is required at the same factor combination to guarantee that n becomes greater than f and the sum of squares due to the pure experimental error can be estimated.

The lack-of-fit may *not be significant* if an appropriate model has been found.

Example 6-3: *Tests for model adequacy*

The straight-line model in Example 6-1 is to be tested for its adequacy. First the F-test for goodness-of-fit is performed according to Eq. 6-19:

$$F(p-1, n-p) = \frac{MSS_{\text{fact}}}{MSS_R} = \frac{6.80}{0.283} = 24.0$$

The critical F-value at the significance level of 0.05 is $F(0.95; 1, 2) = 18.51$ (cf. appendix Table IV). As a consequence the calculated F-value is greater than the critical value and the goodness-of-fit test is indicative of statistical significance.

This result can also be derived from ANOVA-Table 6-3. There the p-value is given as 0.0391. This value is lower than the significance level of 0.05 $(1 - 0.95)$ considered here and, therefore, the test indicates significance. In other words, the risk that all of the factors are different from zero is 3.91%. Since in our model we only have a single factor, x, we can assume with $(100 - 3.91) = 96.09\%$ probability that the effect of x is realistic.

In the second step the F-test for lack-of-fit (Eq. 6-20) is applied:

$$F(f-p, n-f) = \frac{MSS_{\text{lof}}}{MSS_{\text{pe}}} = \frac{0.07}{0.50} = 0.14$$

For the tabulated critical F-value we obtain $F(0.95; 1, 1) = 161$ (cf. appendix Table IV). The F-test for lack-of-fit does not indicate significance, since the calculated F-value is smaller than the critical value. The significance level is only 0.779, as the p-value shows in Table 6-3.

For very small experimental error the F-value in the lack-of-fit test might be extraordinarily large and the test becomes significant. It is to be decided then, whether this result is also *practically significant*.

Confidence intervals

Confidence interval for the parameters

The error in the estimation of the parameters of the straight line in Eq. 6-1 has already been used in the form of the variances for the parameter estimations (Eqs. 6-7 and 6-8).

More generally, the errors of parameter estimation can be calculated on the basis of the variance-covariance matrix (Eq. 5-10). The variance-covariance matrix is computed here on the basis of the mean sum of squares for the pure experimental error as follows:

$$C = MSS_{pe} \left(X^T X\right)^{-1} \tag{6-21}$$

where according to Table 6-1:

$$MSS_{pe} = \frac{SS_{pe}}{n - f}$$

When the sum of squares due to lack of fit, SS_{lof}, is small, the variance of the residuals, s_R^2, can be used instead of the variance due to the experimental error. The residual variance corresponds to the mean sum of squares of the residuals in Table 6-1, i.e.

$$s_R^2 = MSS_R = \frac{SS_R}{n - p}$$

The special case for $p = 2$ we have used already in Eq. 6-9. The variance-covariance matrix is computed then by use of the equation:

$$C = s_R^2 \left(X^T X\right)^{-1} \tag{6-22}$$

The diagonal of the variance-covariance matrix consists of the variances of the parameter estimates and the off-diagonal of those of the related covariances. In the case of the straight-line model with the parameters b_0 and b_1 the corresponding matrix is:

$$C = \begin{pmatrix} s_{b_0}^2 & s_{b_0 b_1}^2 \\ s_{b_1 b_0}^2 & s_{b_1}^2 \end{pmatrix} \tag{6-23}$$

For the confidence interval Δb related to the parameter b we obtain by means of the F-statistic for a given significance level α:

$$b \pm \sqrt{F(\alpha; 1, n - p) s_b^2} \tag{6-24}$$

The covariance matrix (6-21) or (6-22) can be used analogously for calculations of confidence intervals, if more than two parameters are to be determined.

Example 6-4: *Confidence interval*

For the regression parameters of the data in Example 6-1 the confidence intervals of the pararmeters according to Eq. 6-22 must be estimated. The required inversion of the matrix X^TX has already been introduced in Eq. 6-16. With the mean sum of squares of residuals of Table 6-3 we obtain:

$$C = s_R^2 \left(X^TX \right)^{-1} = 0.283 \begin{pmatrix} 0.818 & -0.454 \\ -0.454 & 0.363 \end{pmatrix}$$

$$= \begin{pmatrix} 0.231 & -0.128 \\ -0.128 & 0.103 \end{pmatrix}$$

The standard errors for the parameters of the regression equation are:

$$s_{b_0} = \sqrt{s_{b_0}^2} = \sqrt{0.231} = 0.481$$

$$s_{b_1} = \sqrt{s_{b_1}^2} = \sqrt{0.103} = 0.321$$

The confidence interval of the parameters at a significance level of 0.05 with $F(0.05; 1, 2) = 18.51$ (cf. appendix Table IV) is calculated from:

$$b_0 \pm \sqrt{F(\alpha; 1, n-p) s_{b_0}^2} = 0.41 \pm 2.07$$

$$b_1 \pm \sqrt{F(\alpha; 1, n-p) s_{b_1}^2} = 1.57 \pm 1.38$$

We should be aware that the method given for estimation of the confidence intervals will be valid only if the parameters are *independent* of each other. All elements in the off-diagonals of the covariance matrix in Eq. 6-23 must be zero. For two parameters the confidence intervals describe a square (cf. figure in the margin). If dependences exist between parameters, then for the confidence interval of the parameters an ellipse is obtained. The larger the elements in the off-diagonals in Eq. 6-23, the more pronounced are the deviations from the square shape of the confidence intervals.

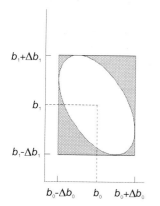

Confidence bands

The previous considerations can also be exploited for computation of the confidence interval for the prediction of a *y*-value, y_0, at given x_0-value. The predicted value and the corresponding confidence interval are given by:

$$y_0 = x_0 b \pm \sqrt{F(\alpha; 1, n-p) s_R^2 \left(1 + x_0 (X^TX)^{-1} x_0^T \right)} \quad (6\text{-}25)$$

For the straight-line model the vector x (dimension $1 \times p$) consists of two elements only (cf. Eq. 6-12), i.e.:

$$x_0 = (1 \quad x_0)$$

The prediction of a single mean from several y-values at a given factor combination is feasible by modifying Eq. 6-25. For m new y-values we obtain for prediction of the mean and its confidence interval:

$$\bar{y}_0 = x_0 b \pm \sqrt{F(\alpha; 1, n-p) \, s_R^2 \left(\frac{1}{m} + x_0 \left(X^T X \right)^{-1} x_0^T \right)}$$

$$(6\text{-}26)$$

For very large values of m Eq. 6-26 simplifies to Eq. 6-27:

$$\hat{y}_0 = x_0 b \pm \sqrt{F(\alpha; 1, n-p) \, s_R^2 \left(x_0 \left(X^T X \right)^{-1} x_0^T \right)} \qquad (6\text{-}27)$$

By means of Eqs. 6-25 to 6-27 the confidence bands for prediction of the y-values along the independent variables can be plotted. This is demonstrated in Fig. 6-3 for the prediction of a single y-value and for a mean from many y-values on the basis of the data in Example 6-1.

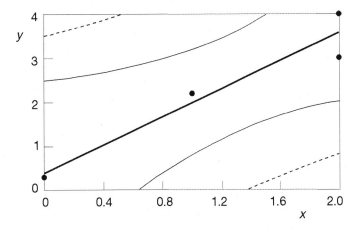

Fig. 6-3.
Confidence bands
for the prediction of individual
y-values (broken lines, Eq. 6-25)
and for the mean from many
y-values (solid lines, Eq. 6-27)
for the data in Table 6-2 of
Example 6-1

Example 6-5: *Confidence interval for prediction of y-values*

On the basis of the straight-line model in Example 6-1

$$y = 0.41 + 1.57x$$

the y-value must be predicted and the corresponding confidence interval estimated for a value of $x = 0.9$. The confidence interval must be given for a single value as well as the mean from several y-values. The variance of the residuals, s_R^2, is 0.283 (cf. Example 6-4). The expression $(X^T X)^{-1}$ has been computed already in Eq. 6-16. As significance level we choose $\alpha = 0.05$, i.e. an F-value of $F(0.05; 1, 2) = 18.51$ (cf. appendix Table IV) must be used.

According to Eq. 6-25 we calculate for the single y-value:

$$y_0 = (1\ 0.9)\begin{pmatrix} 0.41 \\ 1.57 \end{pmatrix} \pm \sqrt{18.51 \cdot 0.283\left[1 + (1\ 0.9)\begin{pmatrix} 0.818 & -0.454 \\ -0.454 & 0.363 \end{pmatrix}\begin{pmatrix} 1 \\ 0.9 \end{pmatrix}\right]}$$

$$y_0 = 1.82 \pm 2.60$$

For prediction of the means from several y-values Eq. 6-27 is applied:

$$\hat{y}_0 = (1\ 0.9)\begin{pmatrix} 0.41 \\ 1.57 \end{pmatrix} \pm \sqrt{18.51 \cdot 0.283\left[(1\ 0.9)\begin{pmatrix} 0.818 & -0.454 \\ -0.454 & 0.363 \end{pmatrix}\begin{pmatrix} 1 \\ 0.9 \end{pmatrix}\right]}$$

$$\hat{y}_0 = 1.82 \pm 1.24$$

As expected, for prediction of a y-value as a mean of several observations a narrower confidence interval results.

Residual analysis

For graphical inspection of regression models the analysis of residuals is applicable. The residual, e_i, denotes the difference between the observed value, y_i, and the value estimated by the model, $\hat{y}_i$. For n observations we get:

$$e_i = y_i - \hat{y}_i \quad \text{with} \quad i = 1, n \tag{6-28}$$

For a valid model the residuals describe the random error of the regression model. The straight-line model in Eq. 6-13 is therefore formulated as follows:

$$y = Xb + e \tag{6-29}$$

where e represents the $1 \times n$-dimensional vector of the random error.

By means of residual analysis the assumptions for the linear regression and the deviations from the model can be checked.

The most important prerequisites for linear regression are:

- The error of the independent variable, x, can be neglected, i.e. x is fixed and only the dependent variable, y, is erroneous.
- The y-values are independent of each other and normally distributed.
- The variances of y-values are comparable for all x-values, i.e. the data are homoscedastic.
- The residuals are then independent, normally distributed and homoscedastic.

If the residuals are plotted in a *histogram*, for large n the shape of a normal distribution results. By plotting the dependence of the residuals on the *run order* trends can be discovered (Fig. 6-4a).

The dependence of the residuals on the predictor variable, x, the observed variable, y, or the predicted variable, $\hat{y}$ are also evaluated. The latter possibilities reveal similar information. This will be explained in the following discussion of the plotting of the dependence of the residuals on the independent variable, x.

- For changing variance of the y-values (heteroscedasticity) different types of residual band result (Fig. 6-4b). For treating those data a transformation of y-values or weighted regression is required.
- A missing linear parameter representing the effect of a linear factor is recognized by linearly ascending or descending residuals (Fig. 6-4c).
- Incomplete models might also be recognized as a result of effects of higher order. In Fig. 6-4d this is demonstrated for the lack of a quadratic term. The residuals then take the form of a parabola.

Apart from the analysis of residuals the recognition of outliers and of influential observations is important for selection of a regression model. Those questions will be raised in the discussion of generalized regression diagnostics in Sec. 6.2.4.

Weighted and robust regression

In residual analysis it has already been mentioned that one prerequisite of conventional regression is an error of comparable size in the y-direction for all realizations of x (homoscedasticity). This means that each point on the straight line has the same weight in the regression analysis. For heteroscedastic data the observations will have different weights in the calculation of the straight line. In order to make the regression less sensitive to small deviations from the distributional assumptions, the influence of observations with large residuals has to be weighted down.

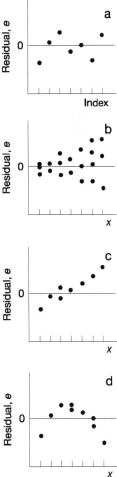

Fig. 6-4.
Residual analysis
in linear regression.
a – time dependent observations;
b – heteroscedasticity;
c – linear effects;
d – quadratic effects

189

Weighted regression

To adjust the weights the analyst can frequently use his expertise. For example, the error in a calibration in the y-direction might increase with increasing x-value, i.e. the relative error is constant. In this case appropriate weights, w_i, would be:

$$w_i = \frac{1}{x_i} \qquad (6\text{-}30)$$

To weight variances changing at the different observation points x_i, a weight based on the reciprocal standard deviation (variance) could be applied:

$$w_i = \frac{1}{s_{y_i}^2} \qquad (6\text{-}31)$$

The weights are usually normalized to unity, i.e. for n observations we obtain:

$$w_i' = \frac{w_i}{\sum\limits_{i=1}^{n} w_i} \qquad (6\text{-}32)$$

The model for the straight line is that given in Eq. 6-1:

$$y = b_0 + b_1 x \qquad (6\text{-}1)$$

By the intercept it is guaranteed that the straight line need not pass through the origin of the coordinate system, but passes through the centroids of the variables, i.e. $\bar{x}$ and $\bar{y}$. For estimation of the parameters for a weighted regression, first the weighted centroids are calculated as follows:

$$\bar{y}_w = \frac{1}{n} \sum_{i=1}^{n} w_i y_i \qquad (6\text{-}33)$$

$$\bar{x}_w = \frac{1}{n} \sum_{i=1}^{n} w_i x_i \qquad (6\text{-}34)$$

The regression parameters are then calculated from:

$$b_1 = \frac{\sum\limits_{i=1}^{n} w_i x_i y_i - n \bar{x}_w \bar{y}_w}{\sum\limits_{i=1}^{n} w_i x_i^2 - n(\bar{x}_w)^2} \qquad (6\text{-}35)$$

$$b_0 = \bar{y}_w - b_1 \bar{x}_w \qquad (6\text{-}36)$$

When predicting x_0-values from p parallel measurements of y_0-values, for the weighted case (cf. Eq. 6-11) the standard deviation is given by:

$$s_0 = \frac{s_{y_w}}{b_1} \sqrt{\frac{1}{p} + \frac{1}{n} + \frac{(\bar{y}_0 - \bar{y}_w)^2}{b_1^2 \left(\sum_{i=1}^{n} w_i x_i^2 - n\bar{x}_w \right)}} \qquad (6\text{-}37)$$

where

$$s_{y_w} = \sqrt{\frac{\left(\sum_{i=1}^{n} w_i y_i^2 - b_1^2 \left(\sum_{i=1}^{n} w_i x_i^2 - n\bar{x}_w^2 \right) \right)}{n-2}} \qquad (6\text{-}38)$$

Standard deviations obtained by weighted regression should be smaller than those obtained by conventional regression. This is advantageous, e.g., for better prediction of y-values.

Example 6-6: *Weighted regression*

The x-y-data in Fig. 6-5 show increasing random error in the y-direction with increasing values of x. To fit those data conventional and weighted regression must be carried out.

By *conventional* regression we obtain for the straight-line regression the following model with the corresponding standard errors of the parameters.

$$y = (1.24 \pm 0.71) + (0.732 \pm 0.091)\, x$$

For weighted regression the reciprocal x-values are used as weights according to Eq. 6-30. The following regression model results:

$$y = (1.08 \pm 0.33) + (0.754 \pm 0.060)\, x$$

The standard errors for estimation of the two parameters of the straight-line model were obviously reduced by applying weighted regression.

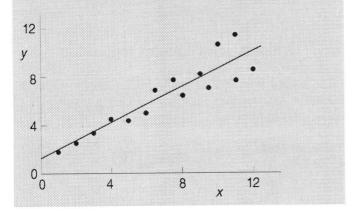

Fig. 6-5.
x-y-values for heteroscedastic data

191

Robustness means tolerance against outliers, i.e. that the model found describes the majority of the data well.

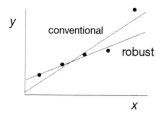

Robust regression

Robust regression is based on iterative weighting of observations. As appropriate weights Tukey suggests:

$$w_i = \begin{cases} \left[1 - \left(\dfrac{e_i}{kS^2}\right)\right] & |e_i| < kS \\ 0 & |e_i| > kS \end{cases} \qquad (6\text{-}39)$$

where e_i is again the residual of observation i; k is a constant that controls the weighting of the residuals and S is a scaling factor, e.g.:

$$S = \text{median}\,\{|e_i|\} \qquad (6\text{-}40)$$

Computation is started by conventional regression analysis. Subsequently the residuals are determined and the weights for a value of $k > 1$ are calculated for each observation. Regression is repeated as long as the parameters change only by a predefined small amount. The smaller the value of k, the more the residuals are weighted down.

6.2 Multiple linear regression

In multivariate modeling several independent and dependent variables may operate. Out of the many regression methods we will learn about the conventional method of ordinary least squares (OLS) as well as methods that are based on biased parameter estimations simultaneously reducing the dimensionality of the regression problem, i.e. **principal** component regression (PCR) and the PLS method (**partial least squares**).

As an example of application of those methods spectrometric multicomponent analysis will be considered. Exemplified by that application regression diagnostics in multiple linear regression are introduced.

6.2.1 Ordinary least squares regression

The general least squares problem that relates a matrix of dependent variables Y to a matrix of independent variables X can be stated as follows:

$$
\begin{pmatrix} y_{11} & y_{12} & \cdots & y_{1m} \\ y_{21} & y_{22} & \cdots & y_{2m} \\ \vdots & & & \\ y_{n1} & y_{n2} & \cdots & y_{nm} \end{pmatrix}
=
\begin{pmatrix} x_{11} & x_{12} & \cdots & x_{1p} \\ x_{21} & x_{22} & \cdots & x_{2p} \\ \vdots & & & \\ x_{n1} & x_{n2} & \cdots & x_{np} \end{pmatrix}
\cdot
\begin{pmatrix} b_{11} & b_{12} & \cdots & b_{1m} \\ b_{21} & b_{22} & \cdots & b_{2m} \\ \vdots & & & \\ b_{p1} & b_{p2} & \cdots & b_{pm} \end{pmatrix}
$$

In matrix notation we get:

$$Y = XB \qquad\qquad (6\text{-}41)$$

where

Y – $n \times m$ matrix of dependent variables
X – $n \times p$ matrix of independent variables
B – $p \times m$ matrix of regression parameters
Residuals – differences between measured and
    predicted data, i.e. $Y - XB$

For example, in multivariate calibration this equation will be used to model the concentrations of m constituents in n samples (Y-matrix) on the n spectra recorded at p wavelengths (X-matrix).

In OLS the number of columns in the X-matrix is maintained. As an example, the set of linear algebraic equations for the first column of Eq. 6-41 looks like this:

$$y_{11} = b_{11}x_{11} + b_{12}x_{12} + \ldots b_{p1}x_{1p}$$

$$y_{21} = b_{11}x_{21} + b_{12}x_{22} + \ldots b_{p1}x_{2p}$$

$$\vdots$$

$$y_{n1} = b_{11}x_{n1} + b_{12}x_{n2} + \ldots b_{p1}x_{np} \qquad (6\text{-}42)$$

OLS is synonymous with the following terms: least squares regression, linear least squares regression, multiple least squares regression, multivariate least squares regression.

Parameter estimation

Usually the matrix of the independent variables, X, is not square so the regression parameters B have to be estimated by the *generalized inverse*. B is given by:

$$B = (X^T X)^{-1} X^T Y \qquad (6\text{-}43)$$

OLS provides the *best linear unbiased estimator* (BLUE) that has the smallest variance among all linear and unbiased estimators.

In principle this equation could be solved by directly inverting the matrix $X^T X$. This, however, will work only if there are no linear dependences and the system is, in a mathematical sense, well-conditioned. The conditioning of the system is given by the *condition number*:

$$\mathrm{cond}\,(B) = \|B\| \cdot \|B^{-1}\| \qquad (6\text{-}44)$$

where $\|B\|$ is the norm of matrix B.

193

The norm of matrix B is computed as its largest singular value (square root of eigenvalue λ) and the norm of B^{-1} as the reciprocal of the smallest singular value of B, i.e.,

$$\text{cond}(B) = \sqrt{\lambda_{max}} \cdot \frac{1}{\sqrt{\lambda_{min}}} \qquad (6\text{-}45)$$

This definition is valid for exactly determined systems where the number of rows in X is equal to the number of columns, i.e., $n = p$ (cf. Eq. 6-41). In the case of overdetermined systems, with $n > p$, the condition number is obtained from:

$$\text{cond}(B) = \left[\text{cond}(B^T B)\right]^{1/2} \qquad (6\text{-}46)$$

Well-conditioned systems have condition numbers near to unity. A matrix is singular if its condition number is infinite; it is ill-conditioned if its condition number is too large, that is, if its reciprocal approaches the machine's floating point precision.

Linear dependences between rows or columns lead to singularities. This might even happen if there is no exact linear dependency and roundoff errors in the machine render some of the equations linearly dependent.

Solution methods

Typical procedures for solving the OLS problem are Gaussian elimination or Gauss-Jordan elimination. More efficient solutions are based on decomposition of the X-matrix by algorithms, such as LU-decomposition, Housholder reduction, or singular value decomposition (SVD). Here one of the most powerful methods, SVD, will be outlined below (cf. Sec. 5.2.1 and 6.2.2).

Significance of parameters

To test for the significance of the parameters of a model we have used already the F-test for goodness-of-fit and for lack-of-fit (Sec. 6.1). In this type of test a whole set of parameters is investigated.

Too many parameters in a model lead to *overfitting* of the experimental observations.

In statistical programs also the test for *individual parameters* on the basis of the t- or F-statistics can be found. The null hypothesis is: the considered parameter differs only randomly from 0. The tests are based on the confidence intervals for the parameters considered, that include the corresponding value $b_i = 0$.

For a t-test on the parameters b_0 of the straight-line model in Eq. 6-1 we get:

$$t = \frac{|b_0 - 0|}{s_{b_0}} \qquad (6\text{-}47)$$

where s_{b_0} is the standard error for the parameter (Eq. 6-7). The calculated t-value is then compared with the critical value at the predefined significance level as discussed in Sec. 2.2.

As long as the model consists of only a single parameter, this kind of testing is adequate. With several parameters the risk of applying a one-parameter test is no longer just α. The probability of rejecting at least one null hypothesis falsely, if all k null hypotheses were true, is here $[1 - (1 - \alpha)^k]$. This means, for example, if the model has two parameters, the risk at significance level of $\alpha = 0.05$ is, accordingly, $[1 - (1 - 0.05)^2] = 0.0975$ or, for three parameters, $[1 - (1 - 0.05)^3] = 0.1426$.

The F-tests demonstrated in Sec. 6.1 deal with a whole set of parameters. The selected significance level is therefore valid irrespective of the number of parameters in the model. However, one should have in mind with those tests that by rejecting the null hypothesis parameters different from zero cannot be recognized individually.

Prediction

Modeling of analytical relationships by estimating the regression parameters in OLS is one of the objectives. Most often in a second step the model parameters are used to predict unknown x- or y-values from measured y- or x-values, e.g., in multivariate calibration the concentrations are predicted from the recorded spectra.

In regression analysis the dependent variable y is also called a *response* variable and the independent variable x is denoted the *predictor* variable or *regressor*.

Prediction of a single y_0-vector from an x_0-vector is easily performed by use of Eq. 6-41. For predicting an x_0-vector (dimension $1 \times p$) from a y_0-vector $(1 \times m)$ the following expression is used:

$$y_0 = x_0 B$$

For the confidence intervals of the predicted y-values the approaches given for the straight-line model are valid (Eqs. 6-25 to 6-27). Instead of the parameter vector b, the matrix of the parameter estimations B must now be used. For prediction of a single y-value according to Eq. 6-25 we obtain:

$$y_0 = x_0 B \pm \sqrt{F(\alpha; p, n - p) s_R^2 \left(1 + x_0 (X^T X)^{-1} x_0^T\right)} \quad (6\text{-}48)$$

Prediction of an x_0-vector (dimension $1 \times p$) from a y_0-vector ($1 \times m$) is achieved by solving a least-squares problem, i.e.:

$$x_0 = y_0 B^T (BB^T)^{-1} \tag{6-49}$$

To estimate the upper limit of error of prediction the following relationship is valid:

$$\frac{\|\delta x_0\|}{\|x_0\|} = \text{cond}(B) \left(\frac{\|\delta y\|}{\|y\|} + \frac{\|\delta B\|}{\|B\|} \right) \tag{6-50}$$

where $\|\delta x_0\|/\|x_0\|$ is the relative error of prediction, $\|\delta y\|/\|y\|$ is the relative error of measurements y, and $\|\delta B\|/\|B\|$ is the relative error of parameter estimation.

From Eq. 6-50 it is obvious that the prediction error will be small if the error in the dependent variable y and the modeling error can be kept low.

6.2.2 Biased parameter estimations: PCR and PLS

PCR – principal component regression

Orthonormal means a set of vectors that are orthogonal and have their norm equal to unity.

PCR is best performed by means of SVD (singular value decomposition). The SVD method was introduced in Sec. 5.2.1. With this method the matrix X is decomposed into two orthonormal matrices U and V joined by a diagonal matrix W of singular values:

$$X = UWV^T \tag{6-51}$$

In OLS the *pseudo-inverse matrix* is equivalent to the generalized inverse:
$$X^+ = (X^T X)^{-1} X^T$$

Computation of the regression coefficients b vector-wise is carried out by formation of the pseudo-inverse matrix X^+ (Moore-Penrose-matrix) according to:

$$X^+ = V \left(\text{diag}(1/w_{ij}) \, U^T \right) \tag{6-52}$$

$$b = X^+ y \tag{6-53}$$

For full rank, all singular values will be substantially different from zero and the SVD solution is equivalent to that from OLS. However, one often obtains several small singular values because of ill-conditioned systems. Therefore, the main goal of PCR is not to keep all singular values for an exact representation of the Moore-Penrose-matrix but to select a subset of singular values that best guarantees predictions of unknown cases.

PLS – partial least squares regression

Regression of Y on X by OLS and PCR is based on solving the linear equations column-wise with respect to the Y-matrix in order to estimate the regression coefficients in the columns of the matrix B in Eq. 6-41. The decomposition of the X-matrix is performed independently from that of the Y-matrix. A method for using the information from the Y-matrix is the PLS-algorithm as developed by H. Wold and propagated by his son S. Wold. Each PLS latent variable direction of the X-matrix is modified so that the covariance between it and the Y-matrix vector is maximized. The PLS method is based on a bilinear model with respect to the objects and the variables of the X- and Y-matrices. Both the X- and Y-matrices are decomposed into smaller matrices according to the following scheme:

$$X = TP^{\mathrm{T}} + E \tag{6-54}$$

$$Y = TQ^{\mathrm{T}} + F \tag{6-55}$$

where X, Y, n, p, m, d have the same meaning as given in Eq. 6-41; T and U are the $n \times d$ scores matrices containing orthogonal rows; P are the $p \times d$ loadings of the X-matrix; E is the $n \times p$ error (residual) matrix of the X-Matrix; Q is the $m \times d$ loading matrix of the Y-matrix; and F is the $n \times m$ error (residual) matrix for the Y-matrix.

To compute the B-coefficients for the general model of Eq. 6-41 the matrices P, Q and W are required:

$$B = W(P^{\mathrm{T}}W)^{-1}Q^{\mathrm{T}} \tag{6-56}$$

with $W - d \times p$ matrix of PLS-weights.

The meaning and estimation of the weight matrix W can be understood from the PLS algorithm in the following example.

Example 6-7: *PLS algorithm*

The first dimension d (index l) is computed from the column mean vectors of the X- and Y-matrices as follows (for centering cf. Eq. 5-2):

$l = 0$:

$$X = X_{\text{original}} - \bar{x} \tag{6-57}$$

$$Y = Y_{\text{original}} - \bar{y} \tag{6-58}$$

Next the dimensions $l = 1$ to $l = d$ are computed on the basis of a suitable stopping criterion, usually the standard error of prediction due to cross-validation (SEP_{cv}, Eq. 6-68):

Loop for the number of dimensions: $l = l + 1$

The principal components are estimated iteratively, e.g. by use of the NIPALS algorithm. Iteration is halted if the precision of the computer is reached.

Iteration loop for NIPALS:

(1) Use the first column of the actual Y-matrix as a starting vector for the y-score vector u:

$$u = y_1$$

(2) Compute the X-weights:

$$w^{\text{T}} = \frac{u^{\text{T}} X}{u^{\text{T}} u} \tag{6-59}$$

(3) Scale the weights to a vector of length one:

$$w^{\text{T}} = \frac{w^{\text{T}}}{(w^{\text{T}} w)^{1/2}} \tag{6-60}$$

(4) Estimate the scores of the X-matrix:

$$t = X w^{\text{T}} \tag{6-61}$$

(5) Compute the loadings of the Y-matrix:

$$q^{\text{T}} = \frac{t^{\text{T}} Y}{t^{\text{T}} t} \tag{6-62}$$

(6) Generate the y-score-vector u:

$$u = \frac{Y q}{q^{\text{T}} q} \tag{6-63}$$

Compare u(old) with u(new). If $\|u(\text{old}) - u(\text{new})\| < \|u(\text{new})\| \times \text{THRESHOLD}$, convergence is obtained, otherwise iteration is continued at step (1). The threshold can be chosen on the basis of precision of the computer.

(7) Determine the inner relationship in the form of a scalar b:

$$b = \frac{u^T t}{t^T t} \qquad (6\text{-}64)$$

(8) Compute the loadings of the X-matrix:

$$p^T = \frac{t^T X}{t^T t} \qquad (6\text{-}65)$$

(9) Form new residuals for the X- and Y-matrices:

$$E = X - btp \qquad (6\text{-}66)$$

$$F = Y - btq \qquad (6\text{-}67)$$

Compute the SEP_{cv}. If the SEP_{cv} is greater than the actual number of factors then the optimum number of dimensions has been found. Otherwise the next dimension is computed. The SEP_{cv} is obtained from:

$$SEP_{\text{cv}} = \left[\frac{\sum_{j=1}^{m} \sum_{i=1}^{n} (y_{ij}^{\text{estimated}} - y_{ij}^{\text{true}})^2}{n \cdot m} \right] \qquad (6\text{-}68)$$

The B-coefficients are finally computed according to Eq. 6-56, i.e.

$$B = W(P^T W)^{-1} Q^T$$

One can show that the matrix $P^T W$ is an upper *bidiagonal* matrix so that the PLS algorithm is merely a variation of diagonalizing a matrix before its inversion.

To estimate the error of the model there are two possibilities. Either the objects or samples are predicted by *resubstitution* into the model equation Eq. 6-41, which gives an estimate of the standard error of modeling, or the model is built by leaving out objects randomly and predicting the left-out samples from the reduced model. The latter prediction is known as *cross-validation* and reveals the standard error of prediction from cross-validation, SEP_{cv} (cf. Eq. 6-68).

The PLS-algorithm is one of the standard methods used for *two-block* modeling, e.g., for multivariate calibration as given below.

6.2.3 Applications in multicomponent analysis

As an example of multivariate modeling we consider the simultaneous determination of several components in analytical systems of low selectivity (multicomponent analysis). These components can be elements, compounds or chemical/physical properties. By means of multicomponent analysis, constituents of pharmaceutical formulations can be determined in the UV-range, the water and protein content of cereal can be estimated from NIR spectra or chemical elements, and technological parameters of coal are predictable on the basis of IR spectra. The limited selectivity of chemical sensors can also be overcome by applying the principles of multicomponent analysis.

The principles are introduced on the basis of spectrometric multicomponent analysis, since this method dominates applications at present.

Spectrometric multicomponent analysis is based on Beer's law formulated for a single component:

$$A_\lambda = \varepsilon_\lambda dc \tag{6-69}$$

where A_λ is the absorbance at wavelength λ; ε_λ is the molar absorption coefficient at wavelength λ, L mol^{-1} cm^{-1}; d is the cell thickness, cm; and c is the molar concentration, mol L^{-1}.

The absorbances can be normalized to a constant cell thickness, so that a simplified Beer's law results:

$$A_\lambda = k_\lambda c \tag{6-70}$$

where k_λ is the normalized absorption coefficient.

In a multicomponent system it is assumed that the absorbances at a specific wavelength i can be represented by the sum of absorbances of the m individual components according to:

$$A_i = k_{i1}c_1 + k_{i2}c_2 + ... + k_{im}c_m = \sum_{j=1}^{m} k_{ij}c_j \tag{6-71}$$

where A_i is the absorbance at wavelength i; k_{ij} is the normalized absorption coefficient of the jth component at wavelength i; and c_j is the concentration of the jth component.

In multiwavelength spectroscopy the spectra are acquired at p wavelengths, so that either exactly determined linear equation systems $(p = m)$ or overdetermined systems $(p > m)$ emerge, i.e.

$$A_1 = k_{11}c_1 + k_{12}c_2 + ... + k_{1m}c_m$$
$$A_2 = k_{21}c_1 + k_{22}c_2 + ... + k_{2m}c_m$$
$$\vdots$$
$$A_p = k_{p1}c_1 + k_{p2}c_2 + ... + k_{pm}c_m$$

In matrix notation we obtain:

$$a = Kc \qquad (6\text{-}72)$$

where a is a $p \times 1$ – dimensional vector that represents the spectrum, K a $p \times m$ – matrix of normalized molar absorptivities, and c a $m \times 1$ – dimensional vector of concentrations.

Direct calibration method

This method is used if all the absorptivities are known. This implies that the pure component spectra can be measured or can be obtained elsewhere, that there is no interaction between the different components of the sample or between constituents and the solvent, and that no unknown matrix constituents interfere with the determination.

Analysis of unknown samples is based on the sample spectrum, a_0, and the known absorptivities K according to (cf. Eq. 6-49):

$$c_0^T = a_0^T K^{-1} \quad \text{for} \quad m = p \qquad (6\text{-}73)$$

or

$$c_0 = K^T \left(KK^T \right)^{-1} a_0 \quad \text{for} \quad m > p \qquad (6\text{-}74)$$

where c_0 is the vector of predicted concentrations.

Example 6-8: *Direct calibration*

Two constituents in a sample must be determined from their absorbances at either two or three wavelengths. The following K-matrix for the absorptivities (arbitrary units) is given:

Wavelength	Constituent	
	1	2
1	3.00	2.00
2	3.00	4.00
3	2.00	6.00

The absorbance data measured for the sample are:

$$a_0 = \begin{pmatrix} 0.71 \\ 1.09 \\ 1.41 \end{pmatrix}$$

The true concentrations of that sample are:

$$c_0 = \begin{pmatrix} c_1 \\ c_2 \end{pmatrix} = \begin{pmatrix} 0.100 \\ 0.200 \end{pmatrix}$$

Determination at the *two wavelengths* 1 and 2 (exactly determined linear system, $p = m$) reveals:

$$c_0^T = a_0^T K^{-1} = (0.71, 1.09) \begin{bmatrix} 3.0 & 3.0 \\ 2.0 & 4.0 \end{bmatrix} = (0.71, 1.09) \begin{bmatrix} 0.666 & -0.5 \\ -0.333 & 0.5 \end{bmatrix} = (0.110, 0.190)$$

or expressed for the single concentrations:

$$c_1 = 0.110 \quad \text{and} \quad c_2 = 0.190$$

As a result, the relative deviation between estimated and true concentrations, in percent, is found to be:

$$\delta c_1 = \frac{c_1^{\text{predicted}} - c_1^{\text{true}}}{c_1^{\text{true}}} \cdot 100 = \frac{0.110 - 0.100}{0.100} \cdot 100 = 10\%$$

$$\delta c_2 = \frac{c_2^{\text{predicted}} - c_2^{\text{true}}}{c_2^{\text{true}}} \cdot 100 = \frac{0.190 - 0.200}{0.200} \cdot 100 = 5\%$$

This relatively high error can be reduced if all *three wavelengths* are used in the analysis:

$$c_0 = K^T (KK^T)^{-1} a_0 = \begin{bmatrix} 3 & 3 & 2 \\ 2 & 4 & 6 \end{bmatrix} \left(\begin{bmatrix} 3 & 2 \\ 3 & 4 \\ 2 & 6 \end{bmatrix} \begin{bmatrix} 3 & 3 & 2 \\ 2 & 4 & 6 \end{bmatrix} \right)^{-1} \begin{bmatrix} 0.71 \\ 1.09 \\ 1.41 \end{bmatrix}$$

$$= \begin{bmatrix} 3 & 3 & 2 \\ 2 & 4 & 6 \end{bmatrix} \left(\begin{bmatrix} 13 & 17 & 18 \\ 17 & 25 & 30 \\ 18 & 30 & 40 \end{bmatrix} \right)^{-1} \begin{bmatrix} 0.71 \\ 1.09 \\ 1.41 \end{bmatrix}$$

$$\begin{bmatrix} 3 & 3 & 2 \\ 2 & 4 & 6 \end{bmatrix} \left(\begin{bmatrix} 0.125 & 0.0479 & -0.0967 \\ 0.0479 & 0.0209 & -0.0309 \\ -0.0967 & -0.0309 & 0.0889 \end{bmatrix} \right) \begin{bmatrix} 0.71 \\ 1.09 \\ 1.41 \end{bmatrix} = \begin{pmatrix} 0.0995 \\ 0.2002 \end{pmatrix}$$

For the relative deviations from the true concentrations we now obtain:

$$\delta c_1 = \frac{c_1^{\text{predicted}} - c_1^{\text{true}}}{c_1^{\text{true}}} \cdot 100 = \frac{0.0995 - 0.100}{0.100} \cdot 100 = 0.5\%$$

$$\delta c_2 = \frac{c_2^{\text{predicted}} - c_2^{\text{true}}}{c_2^{\text{true}}} \cdot 100 = \frac{0.2002 - 0.200}{0.200} \cdot 100 = 0.1\%$$

Notice that the error has been reduced by adding just one wavelength to the analysis scheme. With higher floating-point-precision the result could be even further improved. Therefore *multiwavelength spectrometry* can be a powerful alternative to systems where, for every component, only a single wavelength is applied.

In practice, some or even all of the above mentioned prerequisites for direct calibration on the basis of OLS regression are often not obeyed. Therefore, more sophisticated calibration procedures have to be carried out on the basis of mixture or multivariate calibration.

Indirect calibration methods

Indirect methods are based on estimating the calibration parameters from calibration mixtures. These methods offer the following advantages:

(a) Interactions between constituents or between constituents and the sample matrix can be accounted for in the calibration. Thus, the validity of Beer's law, i.e., the additivity of spectra for every single component and linear response-concentration relationships, is no longer a prerequisite.

(b) Modeling of the background in a principal component becomes feasible.

(c) Systems of highly correlated spectra can also be used for multicomponent analysis.

The different methods for multivariate calibration differ depending on whether the mathematical model is based on Beer's law, i.e., the spectra are regressed on concentrations as with the K-matrix approach, or on inverse models, where the regression of concentrations on spectra is carried out.

Collinearity of data refers to approximate linear dependence among variables.

K-matrix approach

The K-matrix approach is based on an extention of Eq. 6-72 to matrix form:

$$
\begin{bmatrix}
a_{11} & a_{12} & \cdots & a_{1p} \\
a_{21} & a_{22} & \cdots & ap \\
\vdots & & & \\
a_{n1} & a_{n2} & \cdots & a_{np}
\end{bmatrix}
=
\begin{bmatrix}
c_{11} & c_{12} & \cdots & c_{1m} \\
c_{21} & c_{22} & \cdots & c_{2m} \\
\vdots & & & \\
c_{n1} & c_{n2} & \cdots & c_{nm}
\end{bmatrix}
\begin{bmatrix}
k_{11} & k_{12} & \cdots & k_{1p} \\
k_{21} & & & k_{2p} \\
\vdots & & & \\
k_{m1} & k_{m2} & \cdots & k_{mp}
\end{bmatrix}
$$

or in matrix notation

$$A = CK \qquad\qquad (6\text{-}75)$$

where A is the $n \times p$ matrix of absorbances, C is the $n \times m$ matrix of concentrations of constituents, K is the $m \times p$ matrix of absorptivities, n is the number of samples, p is the number of wavelengths, and m is the mumber of components.

In the present notation it is assumed that the absorbance data are centered and that, therefore, there is no intercept at the absorbance axis. If uncentered data are used the first column in the concentration matrix should consist of 1s, and in the K-matrix the intercept coefficients would have to be introduced as the first row.

Calibration is based on a set of n samples of known concentrations for which the spectra are measured. By means of the calibration sample set, estimation of absorptivities is possible by solving for the matrix K according to the general least squares solution:

$$K = (C^{\mathrm{T}}C)^{-1} C^{\mathrm{T}}A \qquad\qquad (6\text{-}76)$$

The *analysis* is then based on the spectrum a_0 $(1 \times p)$ of the unknown sample by use of:

$$c_0 = a_0 K^{\mathrm{T}} (KK^{\mathrm{T}})^{-1} \qquad\qquad (6\text{-}77)$$

where c_0 is the $(1 \times m)$ vector of sought-for concentrations.

A great advantage of the K-matrix approach is the fact that the elements of the K-matrix represent genuine absorptivities with reference to the spectra of the individual constituents. Also, the general assumption in least squares regression analysis is valid, such that only the dependent variable, here the absorbance, is error prone.

In the K-matrix approach, all absorbing constituents of a sample must be explicitly known to be included in the calibration procedure. As we will see below, with more soft modeling techniques it will also be possible to account for unknown constituents without their explicit calibration.

Another disadvantage of the K-matrix approach is that calibration *and* analysis are connected to the inversion of a matrix. Although this is not a problem from the point of view of computational time, it might become a problem if ill-conditioned (less selective) systems are applied, where the spectra of the constituents are very similar. Then in the analysis step (Eq. 6-77) the badly conditioned matrix of absorptivities that must be inverted might be almost singular, i.e. all singular values or eigenvalues are zero. To overcome this difficulty, powerful algorithms for solving the linear equations, such as SVD, should be used in connection with reduction of the dimensionality of the problem.

An alternative to the K-matrix approach is to calibrate the concentrations directly on the spectra. These methods are called inverse calibration methods.

Inverse calibration methods

P-matrix approach

The P-matrix approach is based on the following model:

$$
\begin{bmatrix} c_{11} & c_{12} & \cdots & c_{1m} \\ c_{21} & c_{22} & \cdots & c_{2m} \\ \vdots & & & \\ c_{n1} & c_{n2} & \cdots & c_{nm} \end{bmatrix} = \begin{bmatrix} a_{11} & a_{12} & \cdots & a_{1p} \\ a_{21} & a_{22} & \cdots & a_{2p} \\ \vdots & & & \\ a_{n1} & a_{n2} & \cdots & a_{np} \end{bmatrix} \begin{bmatrix} p_{11} & p_{12} & \cdots & p_{1m} \\ p_{21} & p_{22} & & p_{2m} \\ \vdots & & & \\ p_{p1} & p_{p2} & \cdots & p_{pm} \end{bmatrix}
$$

and in matrix notation:

$$
C = AP \tag{6-78}
$$

The calibration coefficients are now the elements of the P-matrix that are estimated by the generalized least squares solution (OLS) according to:

$$
P = (A^T A)^{-1} A^T C \tag{6-79}
$$

Analysis is carried out by direct multiplication of the measured sample spectrum a_0 by the P-matrix:

$$
c_0 = a_0 P \tag{6-80}
$$

A disadvantage of this calibration method is that the calibration coefficients (elements of the P-matrix) have no physical meaning, since they do not reflect the spectra of the individual components. The usual assumptions about error-free independent variables (here the absorbances) and error-prone dependent variables (here concentrations) are not valid. Therefore, if this method of inverse calibration is used in connection with OLS for estimating the P-coefficients there is only a slight advantage over the classical K-matrix approach because a second matrix inversion is avoided. However, in connection with more soft modeling methods, such as PCR or PLS, the inverse calibration approach is one of the most frequently used calibration tools.

Soft modeling

The methods of soft modeling are based on the inverse calibration model where concentrations are regressed on spectral data:

$$C = AB \qquad (6\text{-}81)$$

where C and A are again the $n \times m$ concentration and $n \times p$ absorbance matrices, respectively, and B is the $p \times m$ matrix of regression or B-coefficients.

PCR-approach

The method of PCR was outlined earlier on the basis of singular value decomposition (SVD). For simultaneous spectroscopic multicomponent analysis the decomposition of the absorbance matrix A can be written as (cf. Eq. 5-23):

$$A = UWV^{\mathrm{T}} \qquad (6\text{-}82)$$

Estimation of the matrix of regression coefficients B is performed column-wise by use of:

$$b = A^{+}c \qquad (6\text{-}83)$$

with A^{+} being the pseudo-inverse of the absorbance matrix A (cf. Eq. 6-53).

The main advantages of PCR-calibration are as follows:

(a) Decomposition of the absorbance matrix into smaller orthogonal matrices enables reduction of the dimensionality of the problem in the case of ill-conditioned systems. So, if highly correlated spectra are to be investigated, one will always obtain the best solution even in case of nearly singular matrices.
(b) Additional unknown components or background components can be automatically modeled as principal components if the concentrations of those components vary within the different calibration samples.

Problems might occur if small principal components are eliminated in the process of reducing the number of significant principal components/singular values, because it might happen that one of the eliminated singular values is important for the prediction of a certain constituent concentration. The decomposition of the absorbance matrix A does not consider relationships between the concentrations and the absorbances. Therefore the decomposition might be not optimal with respect to further use of the calibration model for prediction of concentrations in unknown samples.

A method that accounts for the concentration-spectra relationships during decomposition is the PLS approach.

Details of the PLS method were given in earlier in this section. In multicomponent analysis we obtain the following equations for the decomposition of the absorbance matrix A (the former X-matrix) and the concentration matrix C (formerly the Y-matrix) according to the inverse calibration model in Eq. 6-84:

$$C = AB \qquad (6\text{-}84)$$

$$A = TP^\mathrm{T} + E \qquad (6\text{-}85)$$

$$C = UQ^\mathrm{T} + F \qquad (6\text{-}86)$$

$$B = W(P^\mathrm{T}W)^{-1}Q^\mathrm{T} \qquad (6\text{-}87)$$

The meaning of the additional matrices is the same as in Eqs. 6-54 to 6-56.

The main advantage of the PLS method is based on the interrelated decomposition of the concentration matrix C and the absorbance matrix A, so that with this algorithm the most robust calibrations at present can be obtained.

6.2.4 Regression diagnostics

Leading vendors of software for multicomponent analysis nowadays provide a great variety of tools for diagnosing the suitability of the calibration model, for detecting outliers and influential samples or for estimating realistic prediction errors. It is not unusual that for a calibration set consisting of 30 standard samples, about 5000 different diagnostic plots could be generated.

Visual inspection should be possible from plots of predicted versus measured concentrations, from principal component plots of loadings and scores in the case of soft modeling techniques, and by plotting the dependence of the standard error of calibration (*SEC*) or the standard error of prediction (*SEP*cv, Eq. 6-68) from cross-validation on the number of eigenvalues or principal components.

Very important to diagnostic statistics is the study of the residuals. Let us return to the general least-squares model given in Eq. 6-41. We rewrite it here for a single y-variable as follows:

$$y = Xb + e \qquad (6\text{-}88)$$

where e is the vector of residuals of y-values, i.e., the difference between the measured y-value, y, and the y-value estimated by the model, $\hat{y}$; for a single y-value j $e_j = y_j - \hat{y}_j$.

Diagnostic plots in multi-component analysis:

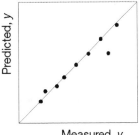

Measured, y

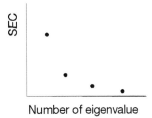

Number of eigenvalue

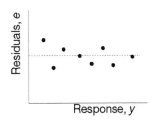

Response, y

With an inverse calibration model one can interpret the model in Eq. 6-88 as regressing the concentrations of a single component, y, on the spectra of the calibration samples collected in the matrix X with n rows (samples) and p columns (wavelength) according to Eq. 6-78. The regression coefficients b are then in a $p \times 1$ vector.

The relationship between the estimated and measured y-values can be described by a fundamental matrix, the *hat-matrix*, H. As explained in Sec. 6.2.1 the regression parameters are estimated by the general inverse as:

$$b = (X^{T}X)^{-1} X^{T} y \tag{6-89}$$

The fitted model has the form:

$$\hat{y} = X\hat{b} \tag{6-90}$$

Substitution of Eq. 6-89 into Eq. 6-90 gives:

$$\hat{y} = X\left[(X^{T}X)^{-1} X^{T} y\right] = X(X^{T}X)^{-1} X^{T} y = Hy \tag{6-91}$$

where H is the $n \times n$ hat-matrix defined by:

$$H = X(X^{T}X)^{-1} X^{T} \tag{6-92}$$

The hat-matrix transforms the vector of measured y-values into the vector of fitted $\hat{y}$-values. The element of H, denoted by h_{ij}, is computed from:

$$h_{ij} = x_{i}^{T} (X^{T}X)^{-1} x_{j} \tag{6-93}$$

Many special relations can be found with the h_{ij}. For example, the *rank* of the matrix X is easily found from the diagonal elements of the hat-matrix by use of the formula:

$$\text{rank}(X) = \sum_{i=1}^{n} h_{ii} \tag{6-94}$$

Further relationships will be learnt subsequently.

Residuals and prediction error

The relationship between the hat-matrix and the residuals can be understood from the following equations:

$$\begin{aligned}
\hat{e} = y - \hat{y} &= y - X(X^{T}X)^{-1} X^{T} y \\
&= \left[I - X(X^{T}X)^{-1} X^{T}\right] = [I - H]y
\end{aligned} \tag{6-95}$$

The residuals depend directly on the product formed by the vector of the measured y-values and the difference between the identity and hat matrix.

Also the elements of the hat matrix are important for estimating the standard error of prediction. In general, the prediction error is calculated as the *predictive residual sums of squares* from:

$$PRESS = \sum_{i=1}^{n} \hat{e}_i^2 = \sum_{i=1}^{n} (y_i - \hat{y}_i)^2 \qquad (6\text{-}96)$$

The influence of observations on a regression model can be assessed by its *leverage*. If the prediction error is to be estimated on the basis of the calibration sample set k samples are left out to calculate new prediction residual errors, $\hat{e}_{(k)}$. By using the elements of the hat-matrix, sometimes also called *leverages* of case k, a good approximation of the estimated prediction error can be obtained from:

$$e_{cv} = \frac{\hat{e}_{(k)}}{1 - h_{kk}} \qquad (6\text{-}97)$$

For the new *PRESS* we obtain:

$$PRESS_{CV} = \sum e_{cv}^2 \qquad (6\text{-}98)$$

On the basis of Eq. 6-98 it is possible to estimate a realistic prediction error without analyzing additional standard samples.

Outliers

Low prediction errors go along with good models. Samples that do not follow the same model as the rest of the data are called *outliers*. Testing for outliers can also be based on the leverage values, h_{ij}. Suppose that the kth sample is an outlier, then a new model is calculated after deleting the kth sample from the data set. On the basis of the new estimate of the regression parameters $\hat{b}_{(k)}$, new residual values, $\hat{e}_{(k)}$, are obtained where the kth sample was not used. To test for the significance of the outlier a Student's t-test can be applied. If $\hat{y}_k$ is not an outlier the null hypothesis can be assumed, such that there is no difference in predicting $\hat{y}_k$ with the full model or with the model estimated without the potential outlier k. If $\hat{y}_k$ is an outlier then the t-value should exceed the critical value at a certain risk level. The t-value can be approximated from the leverage value by use of the formula:

The externally Studentized residual is also termed *jackknifed residual*.

209

$$t_k = \frac{\hat{e}_k}{s_{(k)} \sqrt{1-h_{kk}}} \qquad (6\text{-}99)$$

where t_k is called the externally Studentized residual since case k is not used in computing $s_{(k)}$; it has $n - p - 1$ degrees of freedom; $s_{(k)}$ is derived from the residual mean square, i.e.

$$s_{(k)}^2 = \frac{\displaystyle\sum_{i=1,\, i\neq k}^{n} e_i^2}{n-p-1}.$$

The externally Studentized resiudal is scale invariant. Commonly, an outlier is detected if $t_k > 2$.

Influential observations

Outliers should not be confused with influential observations. Up to now we have used the residuals in order to find problems with a model. If we want to study the robustness of a model to perturbations we do an influence analysis. This kind of study is done as though the model were correct. Influential observations cannot be detected from large residuals. Their removal, however, might cause major changes in the subsequent use of the model. The difference can be understood from Fig. 6-6. A straight line model that includes the influential observation will give a different slope if that observation is deleted. On the other hand if the obvious outlier is included in the model we will estimate larger residuals for all of the cases.

Fig. 6-6.
Straight-line modeling in the presence of an outlier and an influential observation

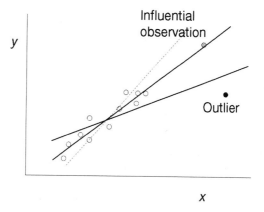

To measure the change due to the influential observation the model has to be built by including or deleting it. From the two models we obtain different estimates of the y-values that can be used to compute a measure, the so-called *Cook's distance* D_k:

$$D_k = \frac{(\hat{\mathbf{y}}_{(k)} - \hat{\mathbf{y}})^{\mathrm{T}}(\hat{\mathbf{y}}_{(k)} - \hat{\mathbf{y}})}{p s^2} \qquad (6\text{-}100)$$

where $\hat{y}_{(k)}$ is the vector of y-values estimated from the model without case k; $\hat{y}$ is the vector of y-values estimated with the full model, and s^2 is the variance computed as in Eq. 6-99 for the full model, i.e.

$$s^2 = \frac{\sum_{i=1}^{n} e_i^2}{n - p}$$

Again the leverage value can be used to estimate the D-value as follows:

$$D_k = \frac{1}{p} r_k^2 \left(\frac{h_{kk}}{1 - h_{kk}} \right) \quad \text{with} \quad r_k = \frac{\hat{e}_k}{s\sqrt{1 - h_{kk}}} \quad (6\text{-}101)$$

where r_k is called the internally Studentized residual because s is estimated by including all of the data.

Large D_k-values reflect the substantial influence of case k. Therefore, samples or cases with the largest D_k-values will be of interest. In practice, those samples should be deleted and the model should be recomputed in order to understand which changes will happen.

Example 6-9: *NIR-Multicomponent analysis of protein in wheat*

As an example of spectroscopic multicomponent analysis the protein content of whole wheat is calibrated on the basis of NIR spectra. The model is then used for analysis of an unkown sample.

In total $n = 30$ calibration samples are available. The NIR spectrum of one sample is given in Fig. 6-7. We will use here the inverse calibration method (cf. Eq. 6-78), i.e. according to the general equation of multiple regression (Eq. 6-41) the protein content is arranged in the y-vector and the matrix X contains the NIR spectra. In order to keep the calibration model small, only the five most important wavelengths are evaluated. The regression model on the basis of OLS reads as follows:

$$y = 6.23 + 670.9 x_1 - 4154 x_2 + 3682 x_3 - 176.8 x_4 - 9.371 x_5$$
$$(6\text{-}102)$$

Table 6-4. ANOVA table for OLS calibration of NIR spectra in the analysis of protein in wheat

Source of variation	SS	df	MSS	F-value	p-level
Corrected for the mean, SS_{corr}	69.80	29	2.41		
Factors, SS_{fact}	66.60	5	13.32	99.81	0.000
Residuals, SS_R	3.203	24	0.1334		

SS – sum of squares, MSS – mean sum of squares, df- degrees of freedom

Table 6-4 provides the results for the analysis of variance. The sum of squares corrected for the mean is explained by 95.41% (coefficient of determination Eq. 6-17) due to the factors, here the wavelength. The F-test for goodness-of-fit is also given in Table 6-4. Based on a significance level of $\alpha = 0.05$ the goodness-of-fit test indicates significance, since the p-level is smaller than 0.05. This means that the parameters of the linear calibration model are significantly different from zero.

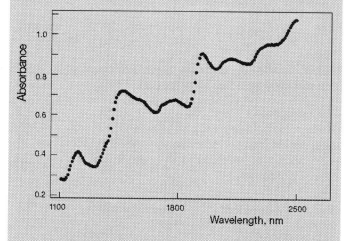

Fig. 6-7.

NIR spectrum of wheat

Fig. 6-8 illustrates the recovery function for the analysis of protein contents in wheat. The predictions correspond here to resubstitution of the calibration samples into the model in Eq. 6-102. For the mean standard error of calibration (SEC) one obtains from the $PRESS$-value (Eq. 6-96):

$$PRESS = \sum_{i=1}^{n} \hat{e}_i^2 = \sum_{i=1}^{n} (y_i - \hat{y}_i)^2 = 3.203$$

$$SEC = \sqrt{\frac{PRESS}{n}} = \sqrt{\frac{3.203}{30}} = 0.327 \qquad (6\text{-}103)$$

To estimate the error for unknown (independent) samples the standard error of prediction of cross validation must be applied (SEP_{cv}, Eq. 6-68). It is expected that this error is greater than for resubstitution and amounts to:

$$SEP_{CV} = \sqrt{\frac{PRESS_{CV}}{n}} = \sqrt{\frac{4.592}{30}} = 0.391$$

Related to the mean protein content of 13 mass% this corresponds to a relative error of $(0.391/13)100\% = 3.0\%$.

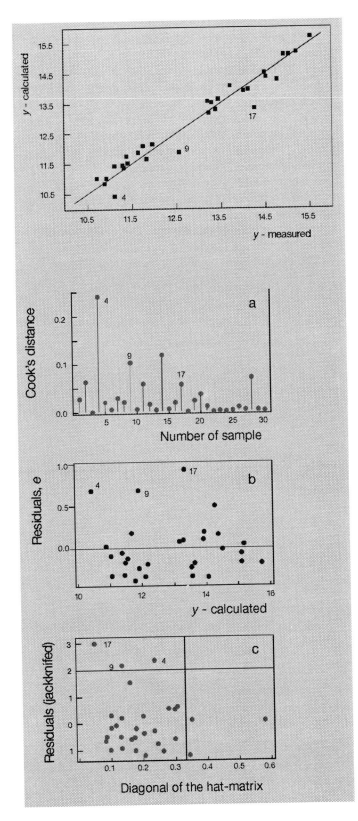

Fig. 6-8.
Recovery function for
resubstitution of samples in the
case of inverse OLS calibration
of protein (mass%) by means of
NIR spectra

Fig. 6-9.
Regression diagnostics
for influential observations
and outliers.
(a) – Cook's distance for
recognition of influential
observations;
(b) – residual plot of dependence
on the calculated y-values;
(c) – jackknifed residuals
according to Eq. 6-99

213

Regression diagnostics are carried out by using Cook's-distance (cf. Fig. 6-9 and Eq. 6-101) and the analysis of residuals. The latter are given as *common residuals* according to Eq. 6-28 and as *jackknifed residuals* after Eq. 6-99 in dependence on the diagonal elements of the hat-matrix. The plot of *Cook's-distances* (Fig. 6-9a) reveals sample 4 as a potential influential observation. However, this sample also has a high residual error, so the model must be recalculated after elimination of the suspicious sample.

A typical outlier is sample number 17. The sample has a very large residual, but cannot be identified as an influential observation.

For computation of the content of the unknown sample the absorbances are inserted into the calibration model (Eq. 6-102). The following protein content in mass% results:

$$y = 6.23 + 670.9 \cdot 0.4569 - 4154 \cdot 0.4178 + 3682 \cdot 0.4134$$
$$- 176.8 \cdot 0.4348 - 9.371 \cdot 0.9816 = 13.29$$

Fig. 6-10.

Error in the calibration of 30 wheat samples by means of the 176 wavelengths if PCR and PLS are used for data evaluation. The *PRESS*-values are computed by use of Eq. 6-96. *PRESS* corresponds to the error due to resubstitution (Eq. 6-96) and $PRESS_{cv}$ to the error estimated by cross-validation (Eq. 6-98)

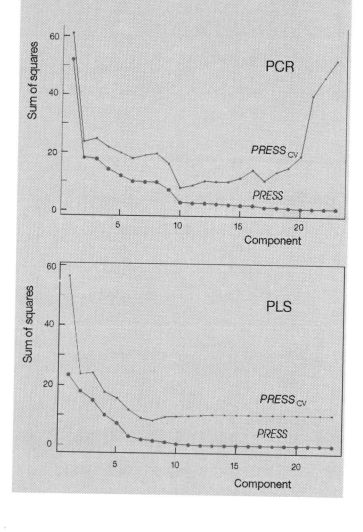

Finally calibration is performed on the basis of the complete spectrum using 176 wavelengths. The OLS method cannot be used for this, since only 30 samples are available and therefore $n < p$, i.e. the matrix is rank-deficient.

The results based on biased parameter estimates, PCR and PLS are represented as their *PRESS*-values in Fig. 6-10.

As expected, the fit of the calibration spectra is improved as the number of components is increased. This effect is here still more pronounced for the PLS method compared with the PCR method. Prediction of unknown samples, however, should not be based on the maximum number of principal components. For estimation of the optimal number of components in respect of robust prediction, the error due to cross-validation, $PRESS_{cv}$, according to Eq. 6-98 is computed on the basis of the *leverage*-value. For both calibration methods we observe a minimum in the dependence on the number of components. It lies for PCR at 10 and for PLS at 8 components.

If the prediction error based on the standard error is compared with computation with only five wavelengths, for both methods, PCR (SEP_{cv} = 0.519) and PLS (SEP_{cv} = 0.526) worse results are obtained. This is because the five optimal wavelengths were selected. The judicious choice of features, here the wavelength, is therefore an important aspect of applying the methods of multicomponent analysis.

6.3   Nonlinear methods

In analytical chemistry nonlinear relationships can frequently be modeled without the application of nonlinear methods. This is feasible by means of *transformations* of variables, such as signals or concentrations. Remember Beer's law in the form

$$I = I_0 e^{-\varepsilon_\lambda dc} \qquad (6\text{-}104)$$

where I is the intensity of transmitted light; I_0 is the intensity of incident light, and ε_λ, d, c have the meanings as given for Eq. 6-69.

The signal intensity is related to the concentration in a nonlinear way. Logarithmic transformation, however, leads to a linear relationship ($A_\lambda = \varepsilon_\lambda dc$, Eq. 6-69), so that the discussed linear methods can be used. The transformation in this example is based on a physical law, i.e. it is of a *mechanistic* nature.

Another possiblity consists in *empirical* transformation on the basis of polynomials of higher order. In this context we have used already quadratic polynomials for response surface methods in Sec. 4.2.3.

Parameters that cannot be transfered by any operation into linear parameters are denoted as being *intrinsically nonlinear*.

In this section on the one hand methods will be introduced that are used to estimate intrinsically nonlinear parameters by means of nonlinear regression analysis. On the other hand we learn about methods that are based on nonparametric, nonlinear modeling. Among those are nonlinear PLS (NPLS), the method of alternating conditional expectations (ACE) and the multivariate adaptive regression splines (MARS).

Nonlinear modeling on the basis of neural networks will be discussed in Sec. 8.2.

6.3.1 Nonlinear regression analysis

Nonlinear parameters can be estimated by the methods of nonlinear regression (NLR). Chemical equilibria represent typical nonlinear models. For example the dependence of retention behavior in high-performance liquid chromatography (HPLC) on the pH-value or the hydrogen ion concentration is described by a set of parameters including distribution coefficients and acid protolysis constants:

$$k' = \frac{k_0 + k_1 \dfrac{[H^+]}{K_{A_1}} + k_{-1} \dfrac{K_{A_2}}{[H^+]}}{1 + \dfrac{[H^+]}{K_{A_1}} + \dfrac{K_{A_2}}{[H^+]}} \qquad (6\text{-}105)$$

where k' is the capacity factor as a relative measure of the retention time; $[H^+]$ is the molar concentration of hydrogen ions; k_0, k_1, k_{-1} are the distribution coefficients for the different forms of the analyte molecules HS, HS^+ and S^-, and K_{A_1}, K_{A_2} are the acid constants for HS and S^-, respectively.

In mathematical terms the nonlinear model in Eq. 6-105 is expressed in terms of the dependent variable y, the independent variable x and the parameter b_i by the following equation:

$$y = \frac{b_0 + b_1 \dfrac{x}{b_3} + b_2 \dfrac{b_4}{x}}{1 + \dfrac{x}{b_3} + \dfrac{b_4}{x}} \qquad (6\text{-}106)$$

In this equation we note two types of parameter: the linear parameters b_0, b_1, and b_2 and the intrinsically nonlinear parameters b_3 and b_4. By appropriate reshaping of the equation the linear parameters can be determined by the methods of linear algebra. In fact there are algorithms for separation of linear and nonlinear parameters in a given model.

The nonlinear parameters can be estimated by means of nonlinear regression. There, however, no closed solution is feasible, but the parameters are iteratively approximated on the basis of their initial values. The following approximation methods can be distinguished:

- The method of steepest descent
- Linearization methods
- Search methods

In all cases the procedures involve minimization of an objective function, γ^2, which represents the deviations of the n estimated y-values from the observed ones:

$$\gamma^2 = \sum_{i=1}^{n} (y_i - \hat{y}_i)^2 \Rightarrow \text{Minimum} \qquad (6\text{-}107)$$

Method of steepest descent

With this method the direction of steepest descent is sought on a plane or hyperplane of the objective function in dependence on the parameters of the model. The basis is a design, e.g. 2^m, in the m parameters where the objective function, γ^2 (Eq. 6-107), is approximated by means of a linear model in the parameters a:

$$\gamma^2 = a_0 + \sum_{j=1}^{m} a_j \frac{[b_j - \bar{b}_j]}{s_j} + e \qquad (6\text{-}108)$$

where $\bar{b}_j$ is the centre of the design and s_j is the standard deviation for the parameters as a normalization factor.

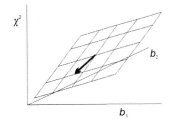

The direction of steepest descent is given by the values $-1 \cdot a_p$, i.e.:

$$\frac{[b_j - \bar{b}_j]}{s_j} = \lambda(-a_j) \qquad (6\text{-}109)$$

or

$$b_j = \bar{b}_j - \lambda a_j s_j \qquad (6\text{-}110)$$

where λ is the descent parameter.

The method is repeated with new parameter values b_j until the minimum is found.

Linearization methods

In contrast to linear parameters nonlinear parameters cannot be represented simply by the product of the matrix of independent variables, X, and the parameter vector, b (cf. Eq. 6-13). If,

217

however, approximate initial values, b_0, are known for the parameter vector, b, then the function $y = f(x)$ can be rewritten such that a linear function emerges that is dependent on the difference between the parameter to be estimated and its starting value, i.e. $\Delta b = b - b_0$. This means, the method can be traced back to the solution of a least-squares problem.

The basis for linearization is the extension of the function into a Taylor series and truncation after the first-order term under the prerequisite that the difference Δb is small enough:

$$y_i = f(x_i, b_1, b_2, ..., b_m) + e_i \tag{6-111}$$

$$y_i = f(x_i, b = b_0) + \left.\frac{\delta f_i}{\delta b_1}\right|_{b=b_0} \Delta b_1 + \left.\frac{\delta f_i}{\delta b_2}\right|_{b=b_0} \Delta b_2 + ... + \left.\frac{\delta f_i}{\delta b_m}\right|_{b=b_0} \Delta b_m + e_i \tag{6-112}$$

where $\left.\dfrac{\delta f_i}{\delta b_j}\right|_{b=b_0}$ is the first derivative of the parameters b_j at point i at the initial value b_0. Eq. 6-112 is then linear in the parameters Δb_1, Δb_2,... Δb_m and similar to the general equation for linear regression (Eq. 6-13). The matrix of the independent variables, X, now, however, contains the partial first derivatives of the function for the parameters in the form:

$$\begin{pmatrix} \dfrac{\delta f_1}{\delta b_1} & \dfrac{\delta f_1}{\delta b_2} & \cdots & \dfrac{\delta f_1}{\delta b_m} \\ \dfrac{\delta f_2}{\delta b_1} & \dfrac{\delta f_2}{\delta b_2} & \cdots & \dfrac{\delta f_2}{\delta b_m} \\ \vdots & & & \\ \dfrac{\delta f_n}{\delta b_1} & \dfrac{\delta f_n}{\delta b_2} & \cdots & \dfrac{\delta f_n}{\delta b_m} \end{pmatrix} = X = J \tag{6-113}$$

The matrix is termed *Jacobian matrix*, J. The linearized model then reads:

$$\Delta y = J \Delta b + e \tag{6-114}$$

where the vector Δy represents the difference between the measured y-values and the predicted value at position b_0, i.e.

$$\Delta y = \begin{pmatrix} y_1 - \hat{y}_{1,0} \\ y_2 - \hat{y}_{2,0} \\ \vdots \\ y_n - \hat{y}_{n,0} \end{pmatrix} \tag{6-115}$$

Estimation of the parameter vector Δb is carried out by analogy with Eq. 6-14 by:

$$\Delta b = \left(J^T J\right)^{-1} J \Delta y \qquad (6\text{-}116)$$

The vector of the nonlinear parameters is finally obtained from:

$$b = b_0 + \Delta b \qquad (6\text{-}117)$$

The accuracy of parameter estimates depends primarily on good starting values. Bad initial values might lead to slow convergence or oscillations.

Secondly, the quality of parameter estimates is also determined by the appropriateness of series extention without considering higher-order terms. If the first order is not sufficient, divergencies cannot be excluded. In principle Taylor expansion can also be performed by inclusion of the second derivative *(Hess's matrix)*.

The derivation of a mechanistic model can be carried out analytically in contrast with empirical models, which require numerical derivatives.

Search methods

A particular search method, the *simplex method*, has already been introduced in Sec. 4.3. Although it was explained in connection with the optimization of experimental observations, it can be transferred analogously to the optimization of the objective function in Eq. 6-107.

A very simple method is the *grid search*, where each point at the given grid is evaluated and in that way the minimium is found. If necessary the grid is reduced to estimate the parameters locally more precisely.

With the *Monte-Carlo method* the optimum is located by randomly changing of the parameters to be estimated. The best set of parameters is maintained in subsequent computations.

Extrapolations outside the areas that were used for building the model are more critical for nonlinear than for linear methods.

Marquardt algorithm

In practice a combination of the method of steepest descent and the linearization procedure is preferred. The algorithm is based on a proposition by Levenberg, further developed by Marquardt. As long as the parameter estimations are still poor the algorithm operates on the basis of the steepest descent method. In the course of optimization the linearization method is progressively included.

Regression diagnostics

A particular cost criterion of nonlinear modeling is the minimization criterion, γ^2, by itself. If this criterion becomes zero a perfect fit is obtained. Similarly as in linear regression analysis the residuals can be graphically inspected and the recovery function can be evaluated.

More difficult is estimation of errors for the nonlinear parameters, because there is no variance-covariance matrix. Frequently the error estimations are restricted to a locally linear range. In the linearization range the confidence bands for the parameters are then calculated as in the linear case (Eqs. 6-25 to 6-27). An alternative consists in error estimations on the basis of Monte-Carlo simulations or bootstrapping methods (cf. Sec. 8.2).

6.3.2 Nonparametric methods

ACE – alternating conditional expectations

At the beginning of Sec. 6.3 the possibilities of transformations of variables for modeling nonlinear relationships were discussed. In the ACE method these transformations need not be predefined, but are found by means of an algorithm. To understand the ACE method we start from the linear multivariate model for a single dependent variable y and p independent variables x_j (cf. Eq. 6-13)

$$y = b_0 + \sum_{j=1}^{p} b_j x_j + e \tag{6-118}$$

where b_j is the regression parameter.

The model of the ACE method is, by analogy [6-6]:

$$g(y) = \sum_{j=1}^{p} f_j(x_j) + e \tag{6-119}$$

Here $g(y)$ is a transformation of the y-variable and the f_j are transformations of the variables x_j. The transformation functions are smooth, but unconstrained functions of the variables y and x.

The ACE model is stored in the form of p pairs of $(y_i, g(y_i))$ and $(x_{ij}, f(x_{ij}))$. The shape of the transformation function is usually not analytical, such as $\log x$ or e^y. The transformation is obtained on the basis of an optimality criterion, according to which the variance of the error e in Eq. 6-119 is minimized with respect to the variance of the transformed variable y.

The notion ACE (alternating conditional expectations) derives from the algorithm that performs the estimation of the optimum transformations. On the basis of initial estimates of the transformations (g^0 and f^0) the algorithm improves the function g^k and subsequently f^k. If, for example, f^k is given, then an improved version for g is obtained by computation of:

$$g^{k+1}(y) = \frac{E\left[\left.\sum_{j=1}^{p} f_j^k(x_j)\right| y\right]}{\left\|E\left[\left.\sum_{j=1}^{p} f_j^k(x_j)\right| y\right]\right\|} \qquad (6\text{-}120)$$

E denotes here the expectation value and $\|\cdot\| = [E(\cdot)]^{1/2}$.

By analogy an improved estimate of f is computed if the actual estimation g^k for g is:

$$f_j^{k+1}(x_j) = E\left[\left. g^k(y) - \sum_{i \neq j}^{p} f_i^k(x_i)\right| x_j\right] \qquad (6\text{-}121)$$

The algorithm uses conditioned expectation operators in alternating sequence. It converges to the optimum transformations $g^k \rightarrow g$ and $f^k \rightarrow f$.

When evaluating real data sets the algorithm in Eqs. 6-120 and 6-121 merely serves the purpose of providing the mathematical basis, i.e. how the algorithm should perform in principle. The operator of conditional expectations is replaced by an appropriate *smoothing algorithm*.

The actual transformations can be graphically evaluated. This is shown in Fig. 6-11 for the data from Example 6-10. Linear relationships are also reflected by linear transformations.

Prediction of y-values is carried out in two steps. First the transformations $f(x)$ on the basis of the x-values are obtained from a look-up table. Usually interpolation between two points will be necessary. After that the p functional values are inserted into Eq. 6-119 and the y-value is calculated by adding the individual functions.

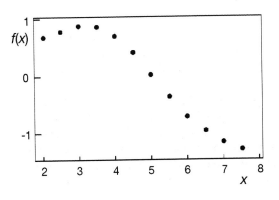

Fig. 6-11.
Plot of the transformation of an x-variable in the ACE method. The plot is based on the data in Example 6-10, i.e. the x-variable corresponds here to a pH-value

For predictions the ACE model can be used until this stage only if the y-values are not transformed, i.e. if $g(y) = y$.

The main advantage of the ACE algorithm is the diversity of possible transformations. For collinear data a previous reduction of dimensionality of the X-matrix, e.g. by means of principal component analysis, is recommended.

NPLS – nonlinear PLS

The linear PLS method was introduced in Sec. 6.2.2, and in connection with multicomponent analysis in Sec. 6.2.3. The NPLS method fits a nonlinear regression model to the data. This model is based on a nonlinear inner relationship of the PLS algorithm, that is described by means of a smoothing function [6-7]. The inner relationship denotes regression of the scores of the Y-matrix on the scores of the X-matrix. According to the algorithm in Example 6-7 (step 7) we obtain, after rearranging for a given dimension, the *inner relationship*:

$$u = f(t) + e \qquad\qquad (6\text{-}122)$$

where t represents the score vector of the X-matrix, u is the score vector of the Y-matrix, e are the residuals, and f is the smoothing function. In the first step the latent variables are computed as in the linear PLS. After that instead of linear approximation of u on t the nonlinear relationship is constructed by means of the smoothing function. The smoothing function is the same as in the ACE method. The nonlinear relationship can be analyzed graphically by plotting the latent variables. In Fig. 6-12 this is demonstrated for univariate data. For a multivariate data set one obtains for each latent x-variable an individual plot.

Fig. 6-12.
Demonstration of the nonlinear relationship between latent variables for the univariate data in Example 6-10

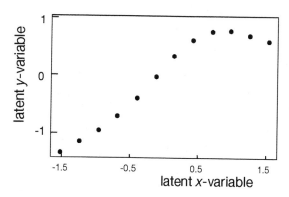

MARS – multivariate adaptive regression splines

MARS is one of the most complex nonlinear methods [6-8]. This method is based on the idea that only a few variables significantly influence the dependent variable in subspaces of multidimensional space. If these subspaces can be identified and, therefore, the corresponding variables assigned, it should be possible to approximate linear or cubic splines to the observations. The fundamentals to *spline functions* were discussed in Sec. 3.1 yet.

The model for a *multivariate regression spline* is based on linear combinations of univariate spline functions S_{kj} of the general form [6-8]:

$$y_i = b_0 + \sum_{k=1}^{K} b_k \prod_{j=1}^{J} \left[S_{kj} (x_{ij} - t_{kj}) \right] + e_i \qquad (6\text{-}123)$$

where b_0, b_k are the parameters of the linear combinations; t_k is the knot at point k; K is the number of knots of the spline, and J is the number of basis functions.

The nonlinearity of the method results from the fact that both the spline coefficients and the linear combinations are optimized with respect to the problem at hand. *Adaptive* means, here, that the subranges are fit to the specific data set. The knots are therefore not fixed as in conventional splines. The search for the most appropriate positions of the knots may be very time-intensive.

Table 6-5.
pH dependence of the retention factor (k') of anthranilic acid in HPLC

pH	k'
2.0	10.60
2.5	14.28
3.0	16.63
3.5	17.60
4.0	15.64
4.5	13.31
5.0	8.09
5.5	4.48
6.0	2.60
6.5	1.92
7.0	1.56
7.5	1.45

Example 6-10: *Nonlinear methods*

The dependence of the retention behavior of anthranilic acid on the pH-value is to be modeled. To solve this problem first the method of nonlinear regression is to be exploited to estimate the parameters according to a physico-chemical model given in Eq. 6-105 and formalized in Eq. 6-106. In a second step the nonparametric methods NPLS and ACE are to be applied.

The measured k'-values are given in Table 6-5 and plotted in Fig. 6-13. Regression by means of NLR provides the parameters summarized in Table 6-6 together with their standard errors.

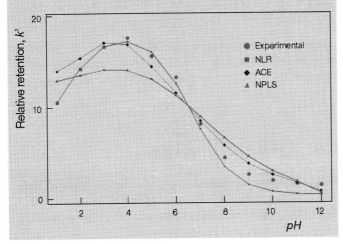

Fig. 6-13.
Modeling of the dependence of the HPLC retention factor (k') of anthranilic acid on the pH-value

223

The parameters can be interpreted in a physico-chemical sense. For example the acid protolysis constants can be compared with those determined by potentiometry. The total error for modeling is given in Table 6-7 as the *PRESS*-value (Eq. 6-96) and the standard error (by analogy with Eq. 6-103).

Table 6-7 also demonstrates the results of modeling with the nonparametric methods NPLS and ACE. The computed retention values are plotted for all three methods in Fig. 6-13. Both plots demonstrate the superiority of the parametric NLR-method. It should be noted that only by regression on the basis of the mechanistic model in Eq. 6-106 can the independent variable be maintained in its antilogarithmic form, i.e. as hydrogen-ion concentration. For the nonparametric computations an acceptable fit of the k'-values is feasible only if pH-values are applied. This is reasoned by the fact that the nonparametric methods can only be used in the sense of curve fitting, but not in the sense of model approximation.

The *standard error* of the parameters b_i is calculated from the diagonal elements of the variance-covariance matrix in Eq. 6-22:

$$\sqrt{s_{b_{ii}}^2}$$

Table 6-6.

Parameters for the model of the retention factor (k') of anthranilic acid in HPLC according to Eq. 6-106

Parameter	Estimation	Standard error
b_0	18.55	0.31
b_1	4.979	1.912
b_2	0.300	0.116
b_3	$7.06 \cdot 10^{-3}$	$2.27 \cdot 10^{-3}$
b_4	$1.52 \cdot 10^{-5}$	$0.11 \cdot 10^{-5}$

Table 6-7.

Error in the modeling of the pH-dependence of the retention (k') of anthranilic acid in HPLC

Method	PRESS	Standard error
NLR	7.647	0.798
NPLS	46.72	1.973
ACE	22.06	1.356

Scientifically a mechanistic model is always to be prefered if it can be constructed. In practice, however, empirical modeling is frequently sufficient, so that nonparametric nonlinear methods are also important tools.

6.4 General reading

[6-1] Henrion, R., Henrion, G., *Multivariate data analysis*, Springer: Berlin, 1994.

[6-2] Draper, N., Smith, H., *Applied Regression Analysis*, 2. Edition, Wiley: New York, 1981.

[6-3]   Box, G. E. P., Draper, N. R., *Empirical Model Building and Response Surfaces*, Wiley: New York, 1987.

[6-4]   Weisberg, S., *Applied Linear Regression*, 2. Edition, Wiley: New York, 1985.

[6-5]   Deming, S. N., Morgan, S. L., *Experimental Design: a Chemometric Approach*, Elsevier: Amsterdam, 1987.

[6-6]   Breiman, L., Friedman, J. H., *Estimating Optimal Transformations for Multiple Regression and Correlation*, J. Am. Statist. Assoc., **80** (1985) 580.

[6-7]   Frank, I. E., *A Nonlinear PLS Model*, Chemometrics Int. Lab. Syst., **8** (1990) 109.

[6-8]   Friedman, J. H., Stuetzle, W., Grosse, E., *Multidimensional Additive Spline Approximation*, Standford University, CA, Technical Report No. 101, August 1988.

[6-9]   Boor, de C., *A Practical Guide to Splines*, Springer: Berlin, 1978.

[6-10] Booksh, K.S., Kowalski, B., Theory of Analytical Chemistry, Anal. Chem. **66** (1994) 782A.

Questions and problems

1. Decide which of the following models contains intrinsically nonlinear parameters which cannot be estimated by linear regression analysis:

a) $y = b_0 + b_1 x + b_{111} x^3$

b) $y = b_0 + \exp(-b_1 x)$

c) $y = b_0 + b_1 \dfrac{1}{x^2}$

d) $y = b_1 x + \dfrac{1}{x + b_2}$

2. The following data were obtained for liquid chromatograms of standard solutions of atrazine. The concentrations are given in mmol L^{-1} and the signals as relative peak area.

Concentration	Area
0.2	0.85
0.2	0.83
0.5	1.34
1.0	2.15
2.0	3.70
3.0	5.15
4.0	6.50
5.0	7.75
6.0	8.90
7.0	9.95

(a) Plot the data and select an appropriate calibration model.

(b) Estimate the parameters of the model by linear regression analysis.

(c) Calculate the confidence intervals of the parameters.

(d) Investigate the residuals of the model and perform a complete ANOVA in order to test the significance of the calibration parameters.

3. Which variations describe the sums of squares in the F tests for 'goodess-of-fit' and 'lack-of-fit' of a model?

4. In which situations are biased regression methods particularly useful compared with ordinary linear regression analysis?

5. Summarize the advantages and disadvantages of direct and inverse multivariate calibration.

6. What is the difference between an outlier and an influential obersvation and how are they detected?

7. Explain the differences between parametric and nonparametric nonlinear regression methods and give examples of their applications.

7 Analytical databanks

Learning objectives

- To introduce the principles of representation, conversion, and storage of analytical data
- To code spectra and chemical structures in analytical data bases and to learn about library search methods and the simulation of spectra

Chemical data bases serve different purposes, such as searching scientific and patent-related literature or retrieving facts about chemical compounds [7-1, 7-2]. In analytical chemistry the data bases that are of interest are those that contain either original measurements (spectra and chromatograms) or derived data, such as concentrations or chemical structures. These data can be retrieved on-line via a network from a host, e.g., STN International. On the other hand, data bases can be stored in individual PCs or in connection with an analytical instrument.

Examples of analytical data bases are given in Table 7-1. Apart from representation of the analytical measurements in the computer the coding of chemical structures is an important aspect of the construction of analytical data bases.

Table 7-1. Some analytical data bases

Method	Data base/Supplier	Data	Compounds	Remarks
NMR	SpecInfo (CC)	120000 spectra	80000	Host/Workstation
	Bruker	19000 spectra		Spectrometer-PC
MS	NIST/EPA/MSDC	50000 spectra	50000	Host/PC
	John Wiley&Sons	120000 spectra	110000	PC
IR	Sadtler	60000 spectra		PC
	Aldrich-Nicolet	> 10000 spectra	10600	Spectrometer-PC
Atomic emission	Plasma 2000 (PE)	50000 atomic lines	60 elements	PC
GC	Sadtler	retention indices		PC

Efficient retrieval of the analytical information depends on appropriate *search strategies*. To confirm a chemical structure on the basis of its spectrum the data base must contain the sought-for spectrum. Very often, however, no spectrum related to the assumed chemical structure is available; then methods for *simulation of spectra* from a chemical structure are needed.

7.1 Representation of analytical information

Type of information

The different sorts of information in analytical data bases include:
- numerical
- alphanumerical, e.g. text
- topological
- graphical

Numerical data are used, if spectral, chromatographic or electroanalytical data have been measured and if concentrations, errors or analysis costs must be stored. Typical alphanumeric data are descriptions of sample identity or analytical procedures. Chemical structures are represented topologically. Graphical information includes plots of spectra or calibration curves as well as data from imaging procedures, e.g., from electron microprobe analysis.

Structure of data bases

The demands for storing and processing information dictates the format of a data base. Usually the content of the data base is acquired in different steps.

Source and library files

The raw analytical data are stored in the source file (cf. Fig. 7-1). Elimination of unimportant data, filtering, transformations or compression of data leads to the library file, which is archived. The library files of an analytical data base consist of a

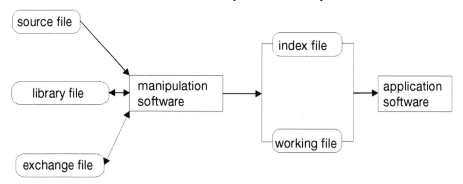

Fig. 7-1.
Structure of analytical data bases

header and a collection of data blocks. The header contains information about the file organization as well as control parameters. Stored in the data block are different sets of data that contain information about the analytical data, such as spectra, about chemical structures and additional remarks.

Exchange files

To transfer data between different bulk storage media, files of fixed exchange formats are created. The most important exchange format in spectroscopy is the JCAMP/DX-format [7-3]. This format has been elaborated by the Joint Committee on Atomic and Molecular Data with the following objectives:

1. Different sorts of spectra should be describable, e.g. FTIR, Raman, UV/VIS, X-ray diffraction, NMR or MS.
2. The text of the file should be readable by computer, humans and telecommunication systems alike. Therefore, only ASCII-characters are allowed.
3. Descriptive information about the sample should be compact.
4. The format must be flexible enough to guarantee later extensions.

Table 7-2 demonstrates an example of a JCAMP/DX data file for storing the IR-spectrum of epichlorhydrin vapour. Here the minimum information is given. Further items concern informa-

Table 7-2.
Important information of a JCAMP/DX exchange file
for an IR-spectrum

1.	##Title= Epichlorhydrin vapor
2.	##JCAMP-DX= 4.24
3.	##DATA TYPE= INFRARED SPECTRUM
4.	##ORIGIN= Sadtler Research Laboratories
5.	##OWNER= EPA/Public Domain
:	
	rows 6-24 are optional
:	
25.	##XUNITS= 1/CM
26.	##YUNITS= ABSORBANCE
27.	##XFACTOR= 1.0
28.	##YFACTOR= 0.001
29.	##FIRSTX= 450.
30.	##LASTX= 4000.
31.	##NPOINTS= 1842
32.	##FIRSTY= 0.058
33.	##XYDATA= (X++(..Y))
34.	450 58 44 34 39 26 24 22 21 21 19 16 15 15 17
35.	etc.
36.	3998 15 15 14
37.	##END=

tion about the compound, e.g. the molecular mass or the Chemical Abstracts number, about sample preparation, the instrument used or about measuring conditions and data-processing methods, such as smoothing or derivatives.

The format is general enough that it can be exploited for similar purposes. Thus, a convention exists for describing chemical structures (JCAMP/CS-format where CS stands for chemical structure).

Coding of spectra

For a long time the limited capacity of bulk storage media hindered the complete storage of spectra and chromatograms in an analytical data base. Therefore, many spectroscopic data bases contain only features of the spectra. For example UV spectra are based on maximum absorbances or IR spectra are represented as peak lists (cf. Fig. 7-2).

Nowadays one attempts to store the complete analytical information, i.e. full spectra, complete chromatograms or even spectro-chromatograms that are generated by hyphenated systems, such as HPLC with diode-array detection. Usually the digitization rate is determined by the measuring conditions. Table 7-3 exemplifies the representation of signal position and intensity of molecular spectra as stored in the beginning of the 1980s in a data base of the ETH Zurich.

As a rule the data bases contain additional information, e.g., about the starting point of a measurement, resolution or multiplets in NMR spectroscopy.

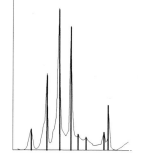

Fig. 7-2.
Evaluation of the peak list
of a spectrum

Table 7-3.
Coding of spectra in the molecular data base of the organic institute of ETH Zurich (OCETH)

Method	Signal position	Intensity	Number of data points
^{13}C NMR	0.1 ppm	0.1%	100
MS	0.1 mass number	0.1%	500
IR	1 cm^{-1} (200...2000 cm^{-1})	%transmission	2300
	4 cm^{-1} (2000...4000 cm^{-1})	at 63 levels	
UV/VIS	1 nm	0.01 in lg ε	400

Apart from the storage of full spectra at certain digitalization levels several algorithms can be used to represent the original data compactly and in an efficient manner for further processing. As an extreme case the measured data or chemical structures are encoded as binary vectors of constant lengths.

As a rule the user does not need to know about the details of data compression. In addition, because of the increasing performance of computers the need to compress the original data decreases, at least for analytical data banks.

Coding of chemical structures

Fragmentation codes

An easy means of converting a chemical structure into a computer-readable format is based on fragmentation codes. Typical fragmentations are aromatic rings, structural skeletons, the OH-group or azo-group (-N=N-). The fragmentations are numbered and stored in a fragmentation list. More generally, a fragmentation is formed on the basis of a freely eligible centre of the molecule and is described by its 1st to 4th spheres by means of hierarchically ordered symbols.

Consider a conventional encoding of chemical structures in ^{13}C NMR spectroscopy as introduced with the so-called HOSE-Code (hierarchically ordered spherical description of environment) [7-4]. Table 7-4 contains some of the symbol descriptions of this code.

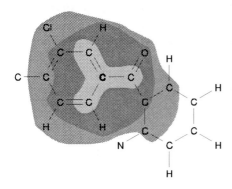

Fig. 7-3.
Spheres around a carbon atom (boldface) as the basis for encoding the structure by means of the HOSE code

Table 7-4.
Symbols for the HOSE-substructure code ordered by priority.

Symbol	Meaning
R	ring
%	triple bond
=	double bond
*	aromatic bond
C	C
O	O
N	N
S	S
X	Cl
Y	Br
&	ring closure
,	separator
(//)	sphere separator

The molecule given in Fig. 7-3 must be coded by the HOSE-code.

The carbon atoms around the centre of the molecule (bold-face C-atom) in the 1st sphere are described according to Table 7-3 by:

*C*CC(

Brackets symbolize the end of the first sphere. The 2nd sphere reads then as:

*C,*C,=OC/

Here the end of the sphere is characterized by the symbol /. Analogously the 3rd and 4th spheres are obtained and the HOSE code for all of the four spheres of the molecule in Fig. 7-3 is:

*C*CC(*C,*C,=OC/*CX,*&,,CC/*&C,,CN,C)

Although with this kind of code a fragmentation can be represented up to the 4th sphere, in many applications the description of the 1st sphere is a good approximation. Rings are characterized by special fragmentation codes, e.g., the HORD code (hierarchically ordered ring description).

In principle, the assignment of fragmentations is easy and unique. Thus, fragmentation codes are also used in structure elucidation in IR spectroscopy, although the vibrations of a molecule are coupled and decomposition of the molecule into individual fragmentations is sometimes misleading for the interpretation process.

Generation of a molecule from its fragments is carried out by means of a *structure generator.*

The easy assignment of fragmentations does not provide the structure of the unknown compound. In a next step the fragmentations must be connected and the hypothetical structure must be compared with candidate structures atom-by-atom. Of course, only those molecules that contain all the found fragmentations can be considered as candidates.

Matrix representation of chemical structures

A mathematical description of a chemical structure can be derived by means of *graph theory.* Fig. 7-4 demonstrates as an example the graph of the molecule phosgene.

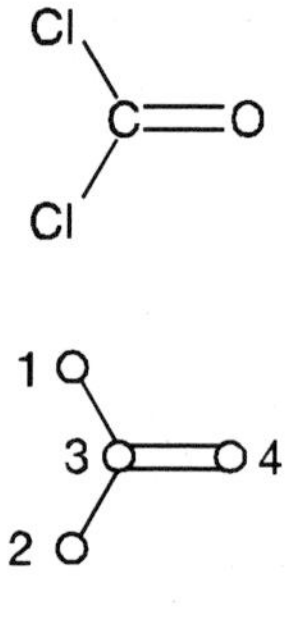

Fig. 7-4.

Chemical structure of phosgene represented as an undirected graph

The atoms of the molecule form the nodes of the graph and the bonds the edges (cf. Table 7-5). The coding of the graph is performed by means of the so-called *adjacency matrix A*. For every element a_{ij} of this matrix the following is valid: if the node K_i is connected with another node K_j, then its value is 1, otherwise it is 0.

Example 7-1: *Adjacency matrix*

For the phosgene molecule we obtain:

$$A = \begin{array}{c} \\ 1 \\ 2 \\ 3 \\ 4 \end{array} \begin{pmatrix} 0 & 0 & 1 & 0 \\ 0 & 0 & 1 & 0 \\ 1 & 1 & 0 & 1 \\ 0 & 0 & 1 & 0 \end{pmatrix} \qquad (7\text{-}1)$$

The additional numbers above and at the side of the brackets correspond to the nodes in the graph as numbered in Fig. 7-4.

Table 7-5.
Representation of a chemical structure based on undirected graphs

Chemical term	Graph-theoretical term	Symbol
molecular formula	molecular graph	G
atom	node	K
covalent bond	edge	g
free electrons	loop	n
topological map	adjacency matrix	A

In the adjacency matrix no bonds are considered. This would be possible by the analogous representation of a molecule based on the *bond-electron matrix*, BE matrix for short. In the latter the *g*-fold connection of two nodes and the number, *n*, of free electrons of an atom is accounted for.

Example 7-2: *Bond-electron matrix*

For the phosgene molecule the BE-matrix *B* is given by:

$$B = \begin{array}{c} \\ 1 \\ 2 \\ 3 \\ 4 \end{array} \begin{pmatrix} 6 & 0 & 1 & 0 \\ 0 & 6 & 1 & 0 \\ 1 & 1 & 0 & 2 \\ 0 & 0 & 2 & 4 \end{pmatrix}$$

$$(7\text{-}2)$$

The advantage of matrix representation consists in the fact that all matrix operations can be applied to the encoded chemical structures. As can be deduced from Eqs. 7-1 and 7-2 quadratic matrices are symmetric around the main diagonal. However, even when the matrices are stored as triangular matrices a disadvantage arises from the need for high storage capacity that accompanies representation of a chemical structure in this way. In addition, the graphs are sparsely connected. Therefore many matrix elements are equal to zero and many parts of the matrix are redundant.

Connection tables

In practice, the node-oriented connection table is applied. This table can be derived from the connection matrix of atoms (cf. Table 7-7) and contains only the numbered chemical elements, the bonds connected to the atoms and the type of the actual bond. Table 7-8 shows the connection table for the phosgene molecule. A less redundant connection table representation is given in Table 7-8.

Table 7-6.
Connection matrix for the molecule phosgene (cf. Fig. 7-4)

	1	2	3	4
1	Cl	0	1	0
2	0	Cl	1	0
3	1	1	C	2
4	0	0	2	O

Table 7-7.
Bond atoms and bonds in a connection table for phosgene
(cf. Fig. 7-4); 1 = single bond, 2 = double bond

Atom number	Atom symbol	Atom1	Bond	Atom2	Bond	Atom3	Bond
1	Cl	3	1				
2	Cl	3	1				
3	C	1	1	2	1	4	2
4	O	3	2				

Table 7-8.
A nonredundant connection table for phosgene (cf. Fig. 7-4);
1 = single bond, 2 = double bond

Node no.	Atom	Connected to	Bond
1	Cl		
2	Cl	3	1
3	C	1	1
4	O	3	2

As a general rule, in the representation of chemical structures the hydrogen atoms are not coded. If necessary, e.g., for graphical representation of a molecule, hydrogen atoms can be added automatically by use of a suitable algorithm.

Connection tables can be easily extended in their rows, e.g., by including information about alternating bonds, cyclic and noncyclic bonds, stereochemistry or by description of variable positions or generic groups in a molecule. Generic groups are represented by Markush structures.

Markush structures

Searching for chemical structures is often related to searching for a whole family of structures rather than for a single compound. The description of general, so-called generic, classes is feasible by means of Markush structures (cf. Fig. 7-5). Such a structure consists of a core with well defined atoms and bonds and at least one general group. The general groups are additionally specified. Substructure search is then performed for a whole compound class. This problem is especially important in the field of patent literature where the most general claim for a compound is envisaged. Eugene A. Markush was the first to apply for a patent in the U.S.A. in 1923 that claimed for the generic class of a chemical compound.

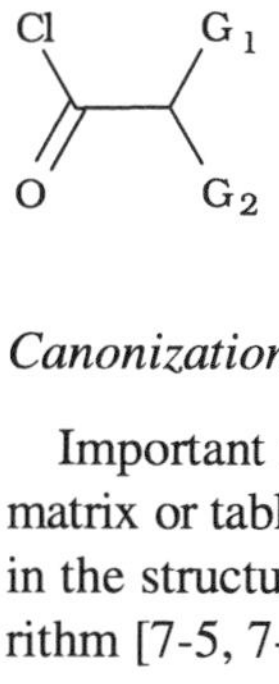

Fig. 7-5.
Example of a Markush structure
with the general groups
G_1 = phenyl, naphthenyl,
N-pyridinyl and G_2 = cycloalkyl

Canonization

Important for the representation of a chemical structure as a matrix or table is the unique assignment (canonization) of atoms in the structure. This can be calculated, e.g. by Morgan's algorithm [7-5, 7-6].

Working and index file

In Fig. 7-1 two additional files are mentioned, i.e., the working and index files. The working file contains the data that are needed in an actual application. Judicious organization of this file guarantees fast access to the data. The organization is determined by the type of information, i.e., whether the data are numerical, alphanumerical, topological or graphical. The searching algorithm also determines the manner in which order the data must be stored.

Access to other data or files is organized via the index file. For processing databank information additional software will be necessary.

LIMS: laboratory-information-
and management-systems

Among analytical data bases there are also systems for organizing laboratory work, for exchanging information and for communicating within a company. These are termed laboratory-information- and management-systems (LIMS).

Typical performance characteristics of a LIMS are:
- Sample identification
- Design of analytical procedures
- Compilation of analytical reports
- Release of analytical results
- Acquisition of raw data and data reduction
- Archiving of analytical results
- Data base functions for chemicals, reference materials, suppliers, specifications, personnel and bibliographies.

Organization of data in a LIMS can be carried out by different models of data bank theory:
- Entity relationship model
- Network data model
- Hierarchical data model
- Relational data model

Today the relational data model is the dominating one. In contrast to a telephone directory this model enables related lists to be represented. It is based on combinations of keys. The individual list can be kept short and the data structure can be extended any time if required. Thus, in the development of a LIMS not all options of the user have to be known in advance.

Table 7-9 shows an example of a relation in a LIMS, where the primary key here serves for sample identification. The origin and matrix of the sample form the attributes of the relation. Every row represents a realization of the relation.

Manipulation of entities in the data base is performed either by relation-oriented operations, such as projection, connection, and selection, or by set-oriented operations, i.e., union, intersection, and negation (cf. Sec. 8.3).

Table 7-9.

Example relation for characterization of an analytical sample

Sample identification	Origin	Matrix
P1	final control	alloy
P2	plant 1	steel
P3	plant 2	fly ash
P4	supplier 4	ore
P5	customer 007	sewage

7.2 Library search

Spectra and chemical structure searches are based on distance and similarity measures as introduced in Sec. 5.2.2. Different strategies are known: sequential search, search based on inverted lists, and hierarchical search trees. The strategies are explained for searching of spectra.

Search strategies

Sequential search

This kind of search is based on comparing the measured spectrum with candidate spectra of the library bit by bit. Sequential search is only useful if a small data set is to be treated or if it is obligatory to retrieve every individual data set. A more efficient way is to sort the entities in a data base by deriving appropriate keys.

Inverted lists

With this method, selected data keys are defined and the data are arranged in new files that contain the information on the individual data sets. Consider an inverted list for a spectral library in the IR-range (Fig. 7-6). The key consists here of the numbered spectral features, i.e., in this example, the wave numbers representing absorption maxima.

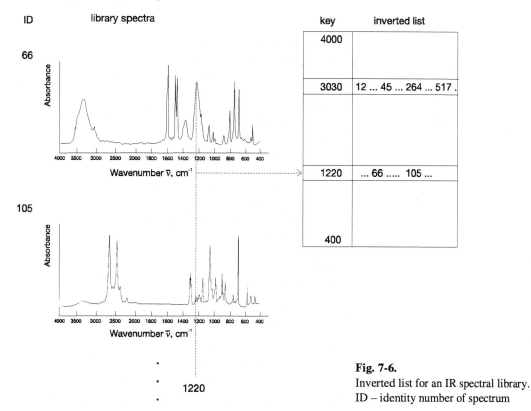

Fig. 7-6.
Inverted list for an IR spectral library.
ID – identity number of spectrum

Every feature appears in the list of keys together with the identity numbers (IDs) of all spectra that contain the actual feature. After collection of all features of the unknown spectrum a rather short file can be generated on the basis of the keys that consist of all the candidate spectra.

Problems may arise if the length of the inverted lists differs. This might be expected because certain wavenumbers are more typical than others or certain chemical structures appear more often, e.g., –C–C– is more frequent than –C=C–. The solution to this problem is the application of Hash coding algorithms: the key is coded by a random number that is then stored in a random access file.

Hierarchical search trees

Hierarchical arrangements of spectra or chemical structures is based on grouping of data by means of some similarity measure. The fundamentals have been introduced with cluster analysis in Sec. 5.2.2. The main problem in library searching is in deciding the metric to be chosen in order to describe the similarity of spectra or chemical structures. In addition, clustering of large amounts of data may still be limited by the computer resources available at present.

Similarity measures for spectra

Comparison of a measured spectrum with a candidate library spectrum is feasible by use of different principles:
- Correlation of spectra
- Similarity and distance measures
- Logical operations

Correlation measures

Comparison of full spectra can be performed successfully by applying correlation measures. The correlation coefficient between the spectra to be compared is computed (cf. Eq. 5-12 in Sec. 5.1). Ranking the comparisons by the size of correlation coefficients provides a *hit list* that describes the quality of the match between the unknown and candidate library spectra (Fig. 7-7). In principle, the spectrum with the highest correlation coefficient is the sought-for spectrum. In order to ensure that a certain degree of similarity is reached for the top spectrum of the hit list a threshold for assigning the library spectrum should be specified.

Typically, the correlation coefficient is used for comparison of UV spectra, e.g., as is common in HPLC with diode-array detection (cf. Table 7-10).

Table 7-10.
Hit list for comparison of UV spectra of corticoides in methanol/water (70%/30%) eluent based on the correlation coefficient (Eq. 5-12 in Sec. 5.1)

Compound	Correlation coefficient
Cortisone	0.999
Dexamethason	0.965
Betamethason	0.962
Prednisolon	0.913

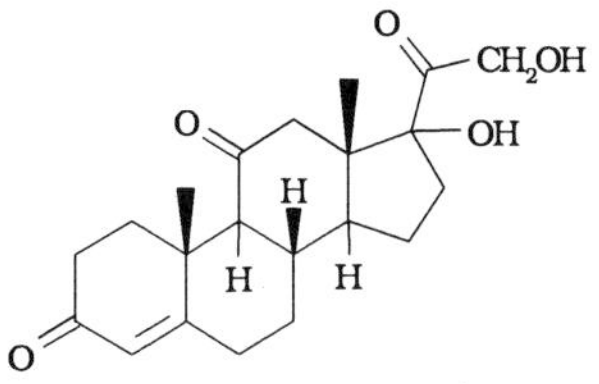

cortisone

Similarity and dissimilarity measures

Comparison of spectra with similarity or distance measures is based on the same definitions as given in Eqs. 5-84 to 5-88 in Sec. 5.2.2.

For full spectra or for other analytical signal curves both the Euclidian distance and the Manhattan distance are used as similarity measures. For the Manhattan distance the differences between the unknown and library spectra are summed. As a result of the comparison, a hit list ranked according to distances or similarities of spectra is again obtained.

Grouping and feature selection

One possibility of speeding up the search is by preliminary sorting of the data sets. Here the methods of unsupervised pattern recognition are used, e.g., principal component and factor analysis, cluster analysis or neural networks (cf. Secs. 5.2.2 and 8.2). The unknown spectrum is then compared with every class separately.

To improve spectral comparisons, selection of features will very often be necessary. For example, in mass spectrometry the original spectra are scarcely used for spectral retrievals. Instead a collection of features is derived, such as the modulo-14-spectrum.

Logical operations

Comparison of spectra is also possible by using logical connectives (cf. truth Table 1-2). A prerequisite for logical comparison is the conversion of spectra into a bit format. Bit-wise conversion can be performed with complete spectra. More frequently, however, the bit vectors are formed from features derived from the raw spectral data. The logical operations can also be considered as distances of the derived vectors in bit space.

In the simplest case the unknown spectrum is compared with the candidate library spectrum by an AND-connective (cf. Table 1-2).

$$0\ 0 \rightarrow 0$$
$$0\ 1 \rightarrow 1$$
$$1\ 0 \rightarrow 1$$
$$1\ 1 \rightarrow 0$$

Fig. 7-7.
Connection of bits by exclusive OR (XOR)

239

A typical dissimilarity measure is the so-called Hamming-distance based on the exclusive OR (cf. Fig. 7-7) calculated as follows:

$$\text{Hamming-distance} = \sum_{i=1}^{p} \text{XOR}(y_i^A - y_i^B) \qquad (7\text{-}1)$$

y_i^A – bit vector for the unknown spectrum at point i

y_i^B – bit vector of the candidate library spectrum

p – number of points (wavelengths)

Figure 7-8 demonstrates the distance calculation. For identical spectra the Hamming-distance would be zero.

Fig. 7-8.
Comparison of an unknown spectrum (A) with a candidate library spectrum (B) by the exclusive OR-connective (XOR)

A   1 0 0 1 1 0 . . .

B   0 1 0 1 1 0 . . .

XOR   1 1 0 0 0 0 . . .

A combination of different logical operations can be found in mass spectrometry with the following dissimilarity measure S′:

$$S' = 2 + \text{XOR} - 2\sum_{i=1}^{p} \text{AND } (y_i^A - y_i^B) \qquad (7\text{-}2)$$

In the discussed examples the comparison is based on bit *vectors*. Another type of logical operation can be based on *set*-oriented comparisons. The latter procedure is necessary if the length of data sets differs from spectrum to spectrum. Typical examples are the peak list in IR spectrometry or capillary gas chromatograms. Fig. 7-9 demonstrates the set-oriented comparison of two spectra.

Fig. 7-9.
Set oriented comparison of two spectra:
A – unknown original spectrum;
B – signal positions for the unknown spectrum;
C – library spectrum with intervals for signal positions;
D – comparison of B and C by intersection of the two sets (AND-connective)

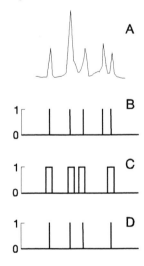

The peak list of the library spectrum is assumed to be error-free, or crisp, and the peak positions of the measured unknown spectrum are characterized by intervals. Comparison is performed usually by the AND-connective, i.e. by the intersection of both sets.

If fuzzy intervals are used the spectra must be compared on the basis of fuzzy set theory (cf. Sec. 8.3).

Similarity measures for chemical structures

If the chemical structures are encoded in *fragmentation codes*, preselection of substructures is feasible. Comparison of the coded vectors of the unknown and library structure is possible by means of AND-connection.

Comparison of structures *atom-by-atom* is based on the connection tables. Consider the classical example presented by E. Meyer in 1970 as given in Fig. 7-10.

Example 7-3: *Substructure search*

The task is to check whether the specified substructure is contained in the given molecule. Starting atom-wise comparison at nitrogen atom no. 1, in both structures the attached atom is carbon. The next atom, however, is oxygen in the molecule and carbon in the substructure. At this point the search is stopped and backtracked. A new trial starting at N atom no. 8 in the molecule will match the substructure step by step.

Fig. 7-10.

Example of substructure search
based on connection tables
of the chemical structures
(for representation of the
connection table cf. Table 7-8)

Molecule:

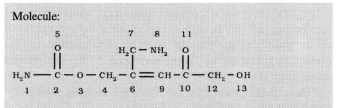

Connection table:

Atom no.	Symbol	Atom1	Bond	Atom2	Bond	Atom3	Bond
1	N	2	1				
2	C	1	1	3	1	5	2
3	O	2	1	4	1		
4	C	3	1	6	1		
5	O	2	2				
6	C	4	1	7	1	9	2
7	C	6	1	8	1		
8	N	7	1				
9	C	6	2	10	1		
10	C	9	1	11	2	12	1
11	O	10	2				
12	C	10	1	13	1		
13	O	12	1				

Sought-for substructure:

$$H_2N \longrightarrow CH_2 \longrightarrow \underset{3}{C} = \underset{4}{CH} - \underset{5}{\overset{\overset{7}{\overset{\displaystyle O}{\|}}}{C}} - \underset{6}{C} -$$

Atom no.	Symbol	Atom1	Bond	Atom2	Bond	Atom3	Bond
1	N	2	1				
2	C	1	1	3	1		
3	C	2	1	4	2		
4	C	3	2	5	1		
5	C	4	1	6	1	7	2
6	C	5	1				
7	O	5	2				

Match
Substructure: 1   2   3   4   5   6
Molecule structure: 8   7   6   9   10   12

7.3 Simulation of spectra

When there is no spectrum in the library which can be used to elucidate a chemical structure, interpretative methods are needed. The methods of pattern recognition and of artificial intelligence must then be used. As a result different chemical structures will be obtained as candidates for the unknown molecule. To verify an assumed structure, simulation of spectra becomes important. In a final step the simulated spectrum could be compared with that measured.

High performing methods for routine simulations of IR and mass spectra are not yet available. In IR spectroscopy the best simulations are obtained on the basis of quantum-chemical approaches.

Simulations are more successful for NMR spectroscopy (Fig. 7-11). Simulation of chemical shifts, δ, are based on increments that are derived from investigations of a set of well-characterized compounds by means of multiple linear regression analysis (cf. Sec. 6.2):

$$\delta = b_0 + b_1 x_1 + b_2 x_2 + ... + b_n x_n \qquad (7\text{-}3)$$

x_i — numerical parameter describing the environment of the structure

b_i — regression coefficient

n — number of descriptors

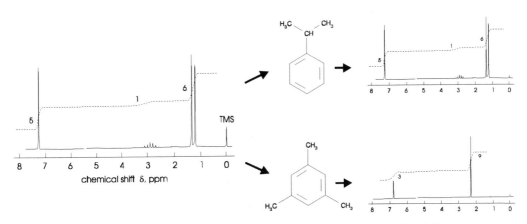

Fig. 7-11. Simulation of the ^{1}H NMR spectrum for cumene (above right) and 1,3,5-trimethylbenzene (below right) to enable verification of the measured spectrum (shown left) as being that of cumene

7.4 General reading

[7-1] Ash, J. E., Hyde, E. (Eds.), *Chemical Information Systems.* Chichester: Ellis Horwood, 1975.

[7-2] Warr, W. A., Suhr, C., *Chemical Information Management.* Weinheim-NewYork: VCH, 1992.

[7-3] McDonald, R. S., Wilks, P. A., *JCAMP-DX: A Standard Form for Exchange of Infrared Spectra in Computer Readable Form.* Applied Spectroscopy, 1988, 42, 151–162.

[7-4] Bremser, W., HOSE – *A Novel Substructure Code*, Anal. Chim. Acta. 1978, 103, 355–365.

[7-5] Morgan, H. L., *The generation of unique machine description for chemical structures – a technique developed at Chemical Abstract Service. J. Chem. Doc.*, 1965, 107–113.

[7-6] Zupan, J., *Algorithms for Chemists.* Chichester: Wiley & Sons, 1989.

Questions and Problems

1. Draw the chemical structure of the molecule given by the following connection table:

Node no.	Atom	Connected to	Bond
1	C		
2	C	1	1
3	C	2	1
4	C	3	2
5	N	3	1

2. Which substructures describe the two fragmentations based on the HOSE-codes = OCC(,*C*C,//) and *C*CC⟹(*C, *C,C//) ?

3. How does one derive a peak list from a full spectrum?

4. Which spectral features can be used as keys for sorting spectra in an inverted list?

5. Mention common similarity measures used for retrieving spectra and chemical structures.

6. Which methods can be used to group spectra and structures?

8 Knowledge processing and soft computing

Learning objectives

- To introduce representation and processing of knowledge in the computer for developing analytical expert systems
- To understand the operation of artificial neural networks and their application for pattern recognition and modeling of analytical data
- To learn about the theory of fuzzy sets for handling vague and incomplete data and knowledge
- To demonstrate the use of genetic algorithms used for solving complex optimization problems

8.1 Artificial intelligence and expert systems

Introduction into AI

A definition of artificial intelligence (AI) is as follows [8-1]:

Artificial intelligence is the study of mental faculties through the use of computational methods.

Among those mental faculties considered are not the especially 'intelligent' human faculties such as abstract thought or reasoning, but rather quite common faculties, such as vision and understanding. Another property of the faculties of interest is that humans can cope with them easily in contrast with tasks a computer can easily perform, e.g. multiplication of 10 digit numbers (cf. Sec. 1.3). Modules of AI research are given in Fig. 8-1.

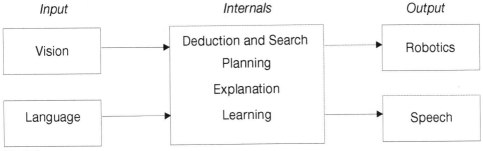

Fig. 8-1.
Modules of AI research by analogy with the mental faculties of human beings

The beginning of *AI research* are ascribed to the Dartmouth Conference in 1956 organized by the scientists John McCarthy and Marvin Minsky. Intensive AI developments were connected in the eighties to a Japanese research project on fifth-generation computers based on the symbolic programming language PROLOG.

The methods of AI are mainly knowledge oriented in contrast with algorithmic data processing considered in previous chapters. Acquisition and processing of knowledge is feasible in two ways:

- First, *symbolically* by means of programming languages, such as LISP or PROLOG. There the knowledge is represented *explicitly* in form of facts and rules. The problem is described rather than solution strategies being implemented as is usual with conventional programming languages.
- Second, *neural networks* are exploited; these may store knowledge *implicitly* and find appropriate answers after presentation of training patterns or structures (cf. Sec. 8.2). Neural networks are built at present by using conventional programming languages. In the future, however, parallel operating computers or transputers will be applied.

Symbolic programming and neural nets are developing increasingly as complementary AI techniques. Neural networks are especially suited for *pattern recognition*. For example images and language can both be interpreted as multidimensional data. Symbolic programming is mainly used for inferencing by logical or approximate reasoning.

Application areas in analytical chemistry

The most important applications of the methods of AI in analytical chemistry are:

- Expert systems
- Intelligent analyzers and robot systems (cf. Sec. 1.3)

The development of *expert systems* is of interest, e.g., for interpretation of spectra. There exist expert systems for both molecular spectroscopy (mass spectrometry, NMR and IR spectroscopy) and atomic spectroscopy, e.g. X-ray fluorescence analysis or atomic emission spectroscopy. In addition in chromatography the selection and optimization of a method can be based on an expert system. Also the selection of analytical methods for a given area is known, e.g. for analyses in a steel laboratory.

Symbolic knowledge processing

Knowledge representation

For internal representation of knowledge in the computer the knowledge has to be preprocessed. A sentence is not stored in its original form, but for example on the basis of *predicate logic*. Consider the following sentence:

'*Copper has an atomic line at 324.8 nm*'

In predicate calculus we write:

atomic line (copper, 324.8 nm)   (8-1)

The predicate is here the term 'atomic line'. The predicate is characterized by the *arguments* for the element 'copper' and the wavelength '324.8 nm'. In general a predicate can be represented by:

predicate (Attribute 1, Attribute 2, ...)   (8-2)

The arguments are instantiated by the different terms in the following way:
- *constant symbols*, such as 'copper', '324.8 nm'
- *variables*, e.g. 'element', 'wavelength'
- *functions* of several predicates.

Complementary forms of knowledge representation are based on *semantic nets* and *frames* (Fig. 8-2). Often they represent just another form of knowledge input. The internal representation is usually based on predicate calculus. The latter can also be interpreted as a relation of objects:

Relation (Object 1, Object 2, ...)   (8-3)

In our example a relation between the element 'copper' and its absorption wavelength at '324.8 nm' is defined, denoted as 'atomic line'.

a

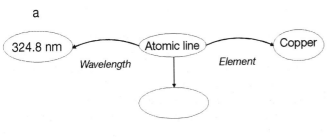

b

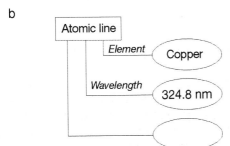

Fig. 8-2.
Representation of knowledge in the forms of semantic nets (a) and frames (b)

Encoding of knowledge by predicate logic or in another computer-internal representation is the task of a knowledge engineer, who is usually a nonexpert in the given knowledge domain.

Inferences

Logical connectives

Inferences consist of *premisses (conditional part)* and consequences *(inference part).*

For reasoning and inferencing on the basis of internal knowledge presentation the relationships or relations must be described. This is feasible with logical connectives, as we know already from Sec. 1.1 (Table 1-2). Important connectives for compound propositions of p and q are:

Conjunction (p AND q)
Alternative (p OR q)
Implication (IF p THEN q)
Negation (NOT p)
Equivalence (p IF AND ONLY IF q)

The connectives already represent a kind of a rule. A *rule* is a conclusion that is true if other conclusions and facts are true. A *fact* is the description of a true proposition, such as in predicate (8-1).

Deduction

The initially known facts are termed *axioms.*

By means of logical connectives correct inferences can be derived in the sense of deductive reasoning. In general for deduction true axioms (postulates) are given and the drawn conclusions are again true. This is denoted a *legal inference.*

> **Example 8-1:** *Deduction*
>
> 'An element is a metal,
>   IF the surface shines,
>   AND the aggregate state is solid,
>   AND the electrical conductivity is high,
>   : '

Abduction

The best known inference rule is the *modus ponens*:
Given:
IF $p = A$ THEN $q = B$
$p = A$
Inference:
$q = B$

With this kind of inference explanations are generated. Example 8-2 demonstrates this.

> **Example 8-2:** *Abduction*
>
> Given: 'X has a high conductivity'
>   'All metals have a high conductivity'
> Inference: 'X is a metal'

248

Abduction is not a legal inference. If in the Example 8-2 X were, e.g., a conducting polymer, then the inference will be false. Apart from this abductive inferences are very useful. Think of a medical or instrumental diagnosis. If certain symptoms are observed, then an assumed disease or instrument state becomes plausible.

Induction

Learning is usually carried out by the third kind of inferencing, i.e. induction. The following example explains this.

Example 8-3: *Induction*

Given: 'Metallic copper is solid'
            'Metallic iron is solid'
                    :

Inference: 'All metals are solid'

Induction is a useful but, as for abduction, is no legal inference. The conclusion drawn in Example 8-3, that 'all metals are solid', is inadmissible, e.g., for the metal mercury.

Search strategies

For systematic search in a knowledge base two strategies are feasible. First in-depth-searching and second in-breadth-searching (Fig. 8-3). With the in-depth-search new states (facts,

In-depth search is also known as *back-tracking*.

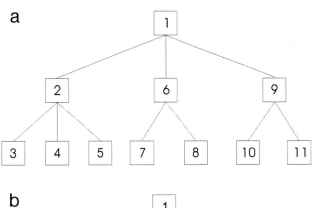

a

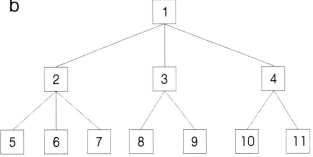

b

Fig. 8-3.
Strategies for in-depth-search (a) and in-breadth-search (b)

rules) are positioned on the top of the logical chain, i.e. the latest retrieved states are eliminated first (last-in/first-out), if they do not satisfy. In contrast to that, in the in-breadth-search the additional states are linked to the end of the chain. The initially tested rules are eliminated first (first-in/first-out).

Symbolic programming

Algorithmic programming languages, such as BASIC, Pascal or C, do not support logical programming. Inferences would have to be performed by means of an additional interpreter that is built externally. This is in principle feasible, but quite laborious.

LISP was developed in the US by John McCarthy.

For logical inferences appropriate symbolic languages were developed. The mother tongue of AI is *LISP* (**list** processing) [8-2]. Dialects of this language are FranzLisp, COMMON-Lisp or MacLisp. The language demands quite large computer memory and a powerful computer, so that real versions are better used on a work station rather than a PC. Facts and rules are arranged in lists (cf. Fig. 8-4). The head of a list characterizes the predicate. The subsequent positions represent the arguments according to formula (8-2).

Fig. 8-4.

Program to find an element in a list of elements by means of LISP and a recursive PROLOG procedure

LISP	PROLOG
(member Element List)	member(Element, [Element| _]).
	member(Element, [_| Restlist]):
	member(Element, Restlist).
Example:	*Example*:
(member 'Fe' (Co Cu Fe))	(member('Fe' ('Co' 'Cu' 'Fe'))
(Fe)	YES

In LISP all inference mechanisms must be implemented by the user. This guarantees that no constraints exist with respect to the type of inferences. A disadvantage, however, is the relatively high workload to program the inference schemes.

A symbolic language in which a particular inference mechanism is already predesigned is PROLOG (**pro**gramming in **log**ic) as developed in the 70s by Alain Colmeraner. In the development of that language an attempt was made to describe the problem to the computer and not to pretend individual steps of the solution. Even if this does not completely work, PROLOG is at present one of the highest-level symbolic programming languages. PROLOG is based on Horn clauses, that

Horn clauses provide *only one* consequence, e.g. IF $p = A$ THEN $q = B$, but not IF $p = A$ THEN $q = B$ **AND** $r = C$.

form a subset of the formal system of predicate logic. A fact or rule in PROLOG corresponds, therefore, to the structure of Eq. 8-1. The inference mechanism is based on the *backtracking* algorithm.

The result of processing rules in a symbolic programming language is not a numerical value, but a truth value, i.e. 'true' or 'false'. For true propositions additional information can be output (cf. Fig. 8-4).

Expert systems

The most important application of methods of knowledge processing on the basis of logical inferences are expert systems.

Expert systems are computer programs, that help to solve problems at an expert level.

They contain knowledge about a particular, limited, problem, but one which originates from the real world, the solution of which requires expert knoweldge.

Aims

The aims in the development of an expert system are manifold:
- To *copy the knowledge of an expert*, in order to make it accessible to less qualified co-workers, to objectivate it or to make it available irrespective of time and conditions.
- To *advise an expert* developing or confirming a diagnosis. In this context rare cases with unusual combinations of facts and findings are of special interest. Routinely returning cases are usually better managed by the expert.
- *Extension* of the activity of an expert and, as a result, of his knowledge.
- *Training* and *education* to learn about a particular expert domain.

The performance of existing expert systems for advising or replacing the expert is rather different. Without doubt is the benefit that is received from developing an expert system with respect to structuring the knowledge domain considered and the complementary collection of important material. Secondly, the training of less qualified personal is very useful.

Development tools

The above mentioned symbolic programming languages *LISP* and *PROLOG* are important tools for developing an expert system. As a rule, however, their use is reserved to the work of a science engineer. The analyst develops his expert system by means of so-called *shells*. These systems provide predefined inference mechanisms and contain an 'empty' knowledge base. The latter must be filled with the domain knowledge of the given field.

conventional programming

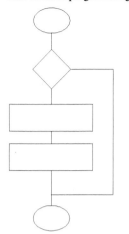

AI program

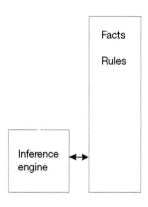

Facts

Rules

Inference engine

Rule
IF → THEN
(Conditional part) (action part)

Structure of expert systems

In conventional programming the different operations are processed in the sense of a spread sheet. In an expert system there is a clear-cut difference between the knowledge base, that contains the facts and rules, and the inference engine. Facts and rules can be easily added to or removed from the knowledge base.

The principle structure of an expert system is given in Fig. 8-5. As a rule, knowledge acquisition is carried out in the dialog between the domain expert, here the analyst, and a knowledge engineer. If the knowledge engineer works in too isolated an environment, the problem emerges that the knowledge base might become too detailed. Redundant information, available as the in-depth knowledge of the analyst, is then maintained. Since, on the other hand, the analyst will not use the tools of AI by himself, in the future the methods for machine learning of symbolic knowledge will become increasingly important.

The operation of the *inference engine* is usually rule-based. Such a system exploits IF-THEN rules (implications). Production systems are termed systems in which a rule interpreter applies the given production rules to a working memory. A PROLOG interpreter is a typical production system.

Apart from the knowledge acquisition component the expert system should also contain an explanation component.

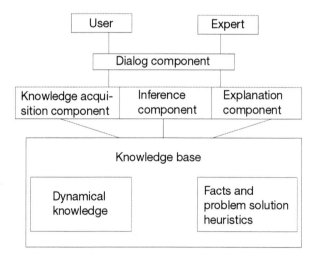

Fig. 8-5.
Structure of an expert system

Expert systems in analytical chemistry

The first expert system – *Dendral* – was developed for interpretation of organic mass spectra [8-3]. A dialect of LISP, i.e. Inter-LISP, was used as the language. Although this system

252

could not be developed as a routinely used expert system, it provided important experience in developing expert systems in analytical chemistry. Perhaps the most intensive studies in the development of expert systems in analytical chemistry were conducted in the framework of the ESPRIT project of the EU [8-5]. Developments from this project are listed together with other expert systems in Table 8-1.

Table 8-1.
Examples of expert systems in analytical chemistry

Area	Aim	Language	Reference
MS	Interpretation of mass spectra	LISP	[8-3]
HPLC	Method selection/optimization	PROLOG, Shells	[8-4, 8-5]
GC	Selection of separation system/operating conditions	LISP	[8-6]
XRA	Spectra interpretation (energy-dispersive)	Pascal	[8-7]
	Spectra interpretation (wavlength-dispersive)	PROLOG	[8-8]
Titration	Karl-Fischer titration	INSIGHT 2-Shell	[8-9]
Steel analysis	Sample preparation/analysis management	Kappa-PC-Shell	[8-10]

8.2 Neural networks

Fundamentals

Application of artifical neural networks for data and knowledge processing is characterized by analogy with a biological neuron. If the firing frequency of a neuron (about 1 kHz) is compared with that of a computer (greater than 100 MHz), then for the neuron this frequency is rather low. The high performance of biological systems therefore is most probably a consequence of the thousand-fold interconnections of about 10 billion cells in the brain.

On the basis of this knowledge one tries to simulate the biological neuron by means of an *artificial neuron*. In the biological neuron the input signal arriving through the axon ends in the synapse. There the information is transformed and sent across the dendrites to the next neuron (Fig. 8-6a). This signal transfer is simulated in the artificial neuron by multiplication of the input signal, x, with the synaptic weight, w, to derive the output signal y (Fig. 8-6b).

What remains in comparison with the biological neuron is a much simplified neuron based on the formation of a simple product. Good performance of artificial neural networks must therefore be attributed to a high degree of interconnections rather than by analogy with a biological neuron.

Synonymous terms for data processing with neural nets are neural computing, parallel distributed processing (PDP) or artificial neural networks.

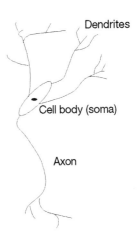

Dendrites

Cell body (soma)

Axon

Fig. 8-6.
Analogy between biological (a)
and an artificial (b) neurons

a

b

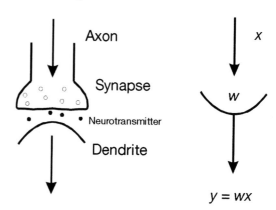

Electrical signal

Axon

Synapse

Neurotransmitter

Dendrite

x

w

$y = wx$

Neurons, weights and transfer functions

The neural networks considered here consist on the one hand of an input layer that receives the input signals. In the simplest case this input layer is connected to a second layer, that represents simultaneously the output layer (two-layer network). Between input and output layer additional layers may be arranged. They are termed *hidden layers* (Fig. 8-7).

Fig. 8-7.
Structure of an artificial
neural network

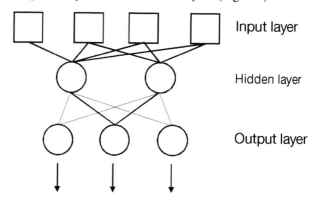

Input layer

Hidden layer

Output layer

Consider the operation of a single neuron j (cf. Fig. 8-8). It receives from n other neurons the input signals, x_i, aggregates them by using the weights of the synapses and passes the result after suitable *transformation* as the output signal y_j.

Fig. 8-8.
Operation of a single neuron

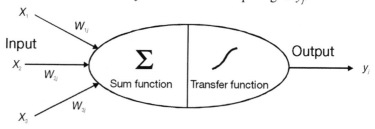

X_1

W_{1j}

Input

X_2

W_{2j}

Σ

Sum function

Transfer function

Output

y_j

W_{3j}

X_3

Typically, for aggregation of the input signal their summation is applied. If the result of the aggregation in neuron j is denoted as the net signal, Net_j, then we obtain:

$$Net_j = \sum_{i=1}^{n} x_i w_{ij} \qquad (8\text{-}4)$$

Other aggregation operators are possible, e.g. formation of the minimum or maximum over all n signals x_i. Before the aggregated signal leaves the neuron, a transformation is performed by means of a transfer function f to furnish the output signal y_j:

$$y_j = f(Net_j) \qquad (8\text{-}5)$$

Important transfer functions are given in Table 8-2. For learning algorithms the derivatives of the transfer functions are needed. They are also given in the table. Apart from linear transfer functions all transformations provide a nonlinear transformation of the aggregated signals. The use of neural networks therefore is especially interesting for solving nonlinear problems.

Transfer functions for threshold logic (A), linear threshold function (B) and sigmoid function (C)

A.
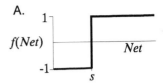

B.
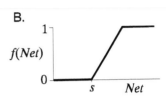

C.
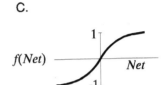

Table 8-2.
Transfer functions for neural nets

Transfer function	Zeroth derivative	First derivative
Threshold logic (binary)	$f(Net) = \begin{cases} 1 & \text{for } Net \geq s \\ 0 & \text{for } Net < s \end{cases}$	
Threshold logic (bipolar)	$f(Net) = \begin{cases} 1 & \text{for } Net \geq s \\ -1 & \text{for } Net < s \end{cases}$	
Linear transfer function	$f(Net) = c \cdot Net$	$f'(Net) = c > 0$
Linear threshold function	$f(Net) = \begin{cases} 1 & \text{for } c \cdot Net \geq s \\ 0 & \text{for } c \cdot Net < s \\ c \cdot Net & \text{else} \end{cases}$	$f'(Net) = c > 0$
Sigmoid function	$f(Net) = \dfrac{1}{1 + e^{-c \cdot Net}}$	$f'(Net) = cf(Net)[1 - f(Net)] > 0$
Hyperbolic-tangent	$f(Net) = \dfrac{e^{c \cdot Net} - c^{-c \cdot Net}}{e^{c \cdot Net} + e^{-c \cdot Net}}$	$f'(Net) = c\left[1 - f(Net)^2\right] > 0$

s – threshold, c – constant

A simple neural net: BAM

The bidirectional associative memory (BAM) is used here to explain the operation of a neural network in more detail.

The BAM consists of an input layer, x, and an output layer, y, as well as of the layer with weights, W. Since the BAM likewise passes and transforms signals in the input and ouput layer, the neurons are simultaneously characterized by circles and squares.

Fig. 8-9.
Bidirectional associative memory consisting of n input and p output neurons

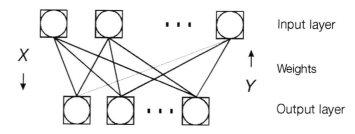

Input layer

Weights

Output layer

To learn m different associations of patterns (x_i, y_j) *correlation learning* is used. The synaptic weights w_{ij} are obtained from:

$$w_{ij} = x_i y_j \qquad (8\text{-}6)$$

The weights for all m associations are stored in a correlation matrix W:

$$W = x_1{}^T y_1 + x_2{}^T y_2 + \ldots + x_m{}^T y_m \qquad (8\text{-}7)$$

If an unknown data vector, x, is to be recognized then it is presented at the input of the net and a first y-vector is estimated:

$$y = xW \qquad (8\text{-}8)$$

The output vector, y, is passed in backward direction by means of:

$$x = yW^T \qquad (8\text{-}9)$$

This procedure is repeated iteratively until a stable state of the network is reached:

$$
\begin{aligned}
&x \rightarrow W \rightarrow y^{(0)} \\
&x^{(1)} \leftarrow W^T \leftarrow y^{(0)} \\
&x^{(1)} \rightarrow W \rightarrow y^{(1)} \\
&x^{(2)} \leftarrow W^T \leftarrow y^{(1)} \\
&\vdots \\
&x^{(f)} \rightarrow W \rightarrow y^{(f)} \\
&x^{(f)} \leftarrow W^T \leftarrow y^{(f)}
\end{aligned}
\qquad (8\text{-}10)
$$

The network takes on a resonant state. Not mentioned is the threshold function applied at each computation of new x- and y-values. Consider the calculations in Example 8-4 in all details.

Example 8-4: *BAM*

For two bipolar associations (x_1, y_1) and (x_2, y_2) the weights of a BAM are to be trained:

$$x_1 = (1 \ -1 \ \ 1 \ -1 \ \ 1 \ -1), \quad y_1 = (1 \ \ 1 \ -1 \ -1) \ (8\text{-}11)$$

$$x_2 = (1 \ \ 1 \ \ 1 \ -1 \ -1 \ -1), \quad y_2 = (1 \ -1 \ \ 1 \ -1) \ (8\text{-}12)$$

After that a new x-vector must be assigned consisting of the following values:

$$x = (-1 \ \ 1 \ \ 1 \ -1 \ -1 \ -1) \tag{8-13}$$

To code the two associations we calculate the matrix of weights (correlation matrix) according to Eq. 8-7:

$$x_1^{\mathrm{T}} y_1 = \begin{pmatrix} 1 \\ -1 \\ 1 \\ -1 \\ 1 \\ -1 \end{pmatrix} (1 \ \ 1 \ -1 \ -1) = \begin{pmatrix} 1 & 1 & -1 & -1 \\ -1 & -1 & 1 & 1 \\ 1 & 1 & -1 & -1 \\ -1 & -1 & 1 & 1 \\ 1 & 1 & -1 & -1 \\ -1 & -1 & 1 & 1 \end{pmatrix}$$

$$x_2^{\mathrm{T}} y_2 = \begin{pmatrix} 1 \\ 1 \\ 1 \\ -1 \\ -1 \\ -1 \end{pmatrix} (1 \ -1 \ \ 1 \ -1) = \begin{pmatrix} 1 & -1 & 1 & -1 \\ 1 & -1 & 1 & -1 \\ 1 & -1 & 1 & -1 \\ -1 & 1 & -1 & 1 \\ -1 & 1 & -1 & 1 \\ -1 & 1 & -1 & 1 \end{pmatrix}$$

$$W = x_1^{\mathrm{T}} y_1 + x_2^{\mathrm{T}} y_2 = \begin{pmatrix} 2 & 0 & 0 & -2 \\ 0 & -2 & 2 & 0 \\ 2 & 0 & 0 & -2 \\ -2 & 0 & 0 & 2 \\ 0 & 2 & -2 & 2 \\ -2 & 0 & 0 & 2 \end{pmatrix}$$

To assign the y-vector associated with the given x-vector in (8-13) multiplication by the weight matrix is carried out (cf. Eq. 8-8):

$$y = xW = (-1 \ \ 1 \ \ 1 \ -1 \ -1 \ -1) \begin{pmatrix} 2 & 0 & 0 & -2 \\ 0 & -2 & 2 & 0 \\ 2 & 0 & 0 & -2 \\ -2 & 0 & 0 & 2 \\ 0 & 2 & -2 & 2 \\ -2 & 0 & 0 & 2 \end{pmatrix} (4 \ -4 \ \ 4 \ -4)$$

Learning paradigms

Neural networks can be used for supervised and unsupervised learning as known from statistical methods of pattern recognition (Chap. 5).

With *unsupervised learning* only input vectors are presented, as known from factorial methods and cluster analysis (cf. Sec. 5.2). The objects are grouped on the basis of their features. Unknown objects can then be automatically recognized. As a typical neural net of that kind we will learn about the Kohonen network below.

For *supervised learning* the network receives in the learning phase additional information about the associated patterns or classifiers that help to adjust the weights. Unknown patterns and classes can be classified subsequently.

With regard to *associations of patterns* we distinguish between auto- and heteroassociations. In the first case the number of input and output neurons is equal. Heteroassociative networks have a different number of neurons in the input and output layers. Pattern associations can be used, e.g., to learn about character or image combinations or spectra-structure relationships.

For *classification* of data vectors the output of the net is coded by the class that corresponds with the input vector. At the end of the network training the unknown data vectors, e.g. spectra, can be assigned on the basis of the value of the output neurons. Without particular constraints, however, a pattern or class is always assigned. This disadvantage we already know from the statistical methods of pattern recognition (Sec. 5.2).

Finally neural networks also can be applied for the purpose of *modeling*. Model parameters are, in a general sense, the weights, w, of the network. Consider the following linear model for a dependent variable y and three independent variables, x_i:

$$y = w_0 + w_1 x_1 + w_2 x_2 + w_3 x_3 \tag{8-14}$$

Neural networks for learning of *heteroassociative* (A) and *autoassociative* (B) *patterns* as well as for *classification* (C)

A.

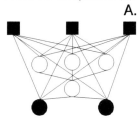

B.

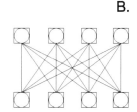

C.

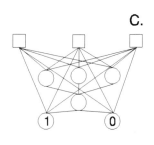

258

This function is mapped in the simplest case by a net consisting of a single layer of weights w_0, w_1, w_2 and w_3. The shift along the ordinate is accounted for by presentation of ones at one neuron, so that the intercept w_0 can be estimated (cf. Eq. 6-1). This particular neuron is termed the *bias neuron*.

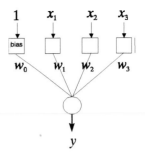

Neural net for *parameter estimation* in Eq. 8-14

Learning laws

Learning in neural networks happens by associative or competitive learning laws. Learning means, in this connection:
- Linking new or elimination of existing synaptic connections.
- Change of the weights of existing connections by minimizing a predefined objective function.

Associative learning laws

The simplest learning law of that kind is the Hebb law. We already have used this rule in the BAM network (Eq. 8-6). According to the Hebb law a weight is strengthened, if the corresponding neurons x_i and y_j are simultaneously excited:

$$w_{ij}(t+1) = w_{ij}(t) + \eta_t x_i y_j \qquad (8\text{-}15)$$

$$\Delta w_{ij} = \eta_t x_i y_j \qquad (8\text{-}16)$$

where η_t is the learning coefficient as a function of time t, e.g.:

$$\eta_t = 0.1\left(1 - \frac{t}{5000}\right)$$

Time t denotes here *cycles*, since the network training is carried out in discrete iterations.

Delta rule

This rule changes the weights in relation to a target pattern t, that is either presented at the input (t_i) or output (t_j) for comparison:

$$\Delta w_{ij} = \eta_t (t_i - x_i) y_j \qquad (8\text{-}17)$$

$$\Delta w_{ij} = \eta_t x_i (t_j - y_j) \qquad (8\text{-}18)$$

A direct application of Hebb's law is given in simple input-output networks. A more general application we will learn about in the backpropagation network in form of the generalized delta rule.

Competitive learning laws

In competitive learning the distance between the input vector, x, and the weight vector, w, is determined by using an appropriate distance measure. Usually the Euclidian distance is applied (cf. Eq. 5-84). In detail the following steps are done:

- Find the winning neuron j. This is the neuron that is nearest to the input vector among all neurons $i = 1, n$:

$$\left\| w_j(t) - x(t) \right\| = \min_i \left\| w_i(t) - x(t) \right\| \qquad (8\text{-}19)$$

where $\| . \|$ is the Euclidian vector norm of x.
- Improve $w_j(t)$ by applying one of the following learning algorithms.

Unsupervised competitive learning

This learning algorithm must be used if no information about the class membership of the training data vectors is available. The change of the weights at iteration t is updated by use of:

$$w_j(t+1) = w_j(t) + \eta_t \left[x(t) - w_j(t) \right] \qquad (8\text{-}20)$$

$$w_i(t+1) = w_i(t) \quad \text{for } i \neq j \qquad (8\text{-}21)$$

If the difference $[x(t) - w_j(t)]$ is greater or lower than zero, the weight of the winning neuron $w_j(t)$ is increased or reduced, respectively. The weights of the other neurons are not changed (Eq. 8-21).

Supervised competitive learning

If class assignments of x-vectors are feasible, the weights can be trained in a supervised fashion. Denote the class membership of the neuron j by D_j, then we have for the algorithm:

$$w_j(t+1) = \begin{cases} w_j(t) + \eta_t \left[x(t) - w_j(t) \right] & \text{for } x \in D_j \\ w_j(t) - \eta_t \left[x(t) - w_j(t) \right] & \text{for } x \notin D_j \end{cases} \qquad (8\text{-}22)$$

The weight vector w_j learns positive if x is correctly classified. It learns negative, or forgets, if x is incorrectly classified. This means, the synaptic changes are performed supervised or 'reinforced'.

Network architecture

For optimization of the architecture of a neural network different possibilities exist:
- Variation of the number of input and output neurons.
- Variation of the number of hidden layers;
- Addition or elimination of neurons in a particular layer;
- Modification of the connections of neurons within a layer and between layers. As a consequence the number of weights is modified;

Competitive learning operates on the basis of a *next neighbor classifier*

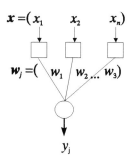

$$x = (x_1 \quad x_2 \quad x_n)$$

$$w_j = (w_1 \quad w_2 \dots w_3)$$

y_j

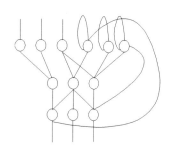

- Selection of those neurons that receive a correction signal;
- Definition of the information flow, that can be directed forward or backward or is recurrent.

The variety of network architectures complicates optimization of a neural network. In addition, further parameters, such as the learning coefficient, must be adjusted. The danger of designing over-dimensioned networks, that are redundant with respect to several parameters, is therefore large.

Neural network models

Perceptron

The simplest neural network is the perceptron. It was introduced by E. Rosenblatt (1950) and served the purpose of optical pattern recognition, i.e. it represents a very simple model of the retina of the eye.

Formally, it can be categorized as a one-layer net with two inputs, x_1 and x_2, and one output, y, in the sense of a classifier for linearly separable classes. For this classifier it can be shown that at time or cycle t:

$$y(t) = f(Net) = f\left(\sum_{i=1}^{n} w_i(t) x_i(t) + s \right) \qquad (8\text{-}23)$$

$$y(t) = \begin{cases} 1 & \text{for } Net \geq s \\ 0 & \text{for } Net < s \end{cases} \qquad (8\text{-}24)$$

Here s is again a threshold.

The learning procedure corresponds to a supervised learning process according to an associative learning law, i.e.

$$w_i(t+1) = w_i(t) + \eta_t \left[D(t) - y(t) \right] x_i(t) \qquad (8\text{-}25)$$

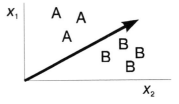

$D(t)$ represents the class membersip:

$$D(t) = \begin{cases} 1 & \text{for } class\,A \\ 0 & \text{for } class\,B \end{cases} \qquad (8\text{-}26)$$

The perceptron can be compared with the linear learning machine (Sec. 5.3.1). As demonstrated by Minsky and Papert (1969), certain problems cannot be solved by using a simple perceptron. As an example the exclusive OR connection must be mentioned as given in Table 8-3 (cf. Fig. 7-8).

Table 8-3.
Exclusive OR (XOR problem)

x_1	x_2	y	class
0	0	0	A
0	1	1	B
1	0	1	B
1	1	0	A

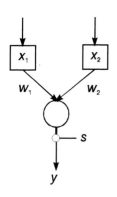

With two input neurons, x_1 and x_2, and the following model for the separation plane it can be shown that:

$$y = f(Net) = \begin{cases} 1 \text{ for } w_1 x_2 + w_2 x_2 \geq s \\ 0 \text{ for } w_1 x_2 + w_2 x_2 < s \end{cases} \tag{8-27}$$

A graphical interpretation of the XOR problem is given in Examples 8-5 and 8-6.

Example 8-5: *XOR problem*

The problem of the exclusive OR is illustrated in Fig. 8-10. The problem is to find a plane that separates the classes A and B from each other. As the separation plane drawn in Fig. 8-10 demonstrates, this is not possible. Also no 'better' separation planes could be found.

Fig. 8-10.
Insolubility of the XOR problem by use of a network based on a *single* separation plane

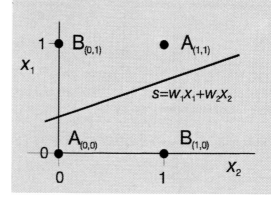

A solution exists if a network is constructed that contains at least two neurons in the hidden layer. Example 8-6 gives one possibility of solving the XOR problem.

Example 8-6: *Solution of the XOR problem*

To solve the XOR problem in Table 8-3 we construct the following neural net with two neurons in the hidden layer:

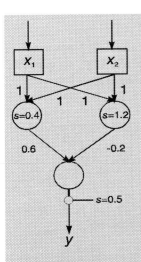

If the weights are adjusted as indicated in the figure above, then the following output values are obtained for the two input vectors in a and b:

a. $x_1 = 1$ and $x_2 = 1$;

$$y = s_{0.5}\left(0.6 \cdot \left(s_{0.4}\left(1 \cdot 1 + 1 \cdot 1\right)\right) - 0.2 \cdot \left(s_{1.2}\left(1 \cdot 1 + 1 \cdot 1\right)\right)\right) =$$

$$s_{0.5}\left(0.6 \cdot \left(s_{0.4}\left(2\right)\right) - 0.2 \cdot \left(s_{1.2}\left(2\right)\right)\right) = s_{0.5}\left(0.6 - 0.2\right) = 0$$

b. $x_1 = 0$ and $x_2 = 1$;

$$y = s_{0.5}\left(0.6 \cdot \left(s_{0.4}\left(1 \cdot 0 + 1 \cdot 1\right)\right) - 0.2 \cdot \left(s_{1.2}\left(1 \cdot 0 + 1 \cdot 1\right)\right)\right) =$$

$$s_{0.5}\left(0.6 \cdot \left(s_{0.4}\left(1\right)\right) - 0.2 \cdot \left(s_{1.2}\left(1\right)\right)\right) = s_{0.5}\left(0.6 - 0\right) = 1$$

Graphical interpretation of the net provides Fig. 8-11.

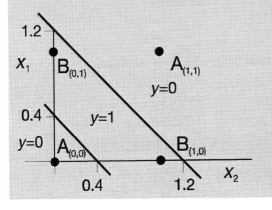

Fig. 8-11.
Solution of the XOR problem with a network that contains two neurons in the hidden layer

The notion *backpropagation* characterizes the kind of error correction done backwards from the output layer:

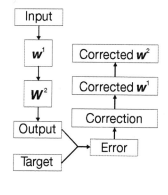

Fig. 8-12.
Multi-layer perceptron as a basis for a backpropagation network

Backpropagation networks

Multi-layer perceptrons (Fig. 8-12) are used nowadays in connection with the backpropagation algorithm. In analytical chemistry more than 90% of the applications are based on this learning algorithm. A first approach towards this forward-directed learning algorithm was made by Werbos (1974). This was further developped by McClelland and Rumelhart (1986).

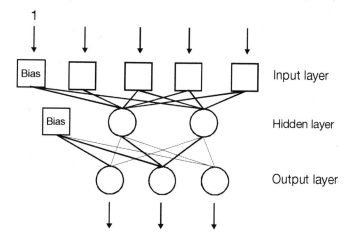

Learning algorithm

The basis of the backpropagation algorithm is the generalized delta-rule (cf. Eq. 8-18). The signal vector produced at the output of the net, y, is compared with a target vector, t. Mathematically the path of steepest descent is traced (cf. Sec. 6.3.1). The individual steps of the algorithm are:

$$\Delta w_{ij}^l (t+1) = \eta_t \delta_j^{l-1} y_i^{l-1} + \alpha \Delta w_{ij}^l (t) \qquad (8\text{-}28)$$

where l is the index of the actual layer, t is time, η_t the learning coefficient, δ the error, and α the momentum.

The momentum is used to increase the convergence rate for learning. By means of a momentum parameter, α, during computation part of the hitherto performed weight changes is maintained.

The error signal in the output layer is obtained from:

$$\delta_j^{\text{last}} = (t_j - y_j) f'(y_j) \qquad (8\text{-}29)$$

For the neurons in the hidden layer the error signal is calculated from:

$$\delta_j^l = \sum_{k=1}^{r} \delta_k^{l+1} w_{kj}^{l+1} f'(y_j^l) \qquad (8\text{-}30)$$

where r is the number of neurons in layer $l + 1$ and the derivation of the sigmoid function is:

$$f'(y) = y(1-y) \qquad (8\text{-}31)$$

Before training of a backpropagation network the following settlements are required:

- Selection of the network architecture including the labeling of the bias-neurons.
- Initialization of weights by random numbers in a predefined interval, e.g. [–0.1, +0.1], or by least-squares estimations.
- Adjusting values for the learning rate and the momentum in Eq. 8-28.
- Selection of a stopping criterion on the basis of a maximum iteration number, a threshold for the error between the output and target value, or of a threshold for the change of weights.

The backpropagation net is an example of a supervised learning neural network. Applications in analytical chemistry are given in Table 8-4. In the following section we consider models of unsupervised learning networks.

Kohonen network

Neural networks for unsupervised learning are based on a competitive layer of weights arranged linearly or in a plane (Fig. 8-13). If arranged in a plane the nets are termed a Koho-

Kohonen networks are also termed *self-organizing nets* or *self-organizing feature maps*.

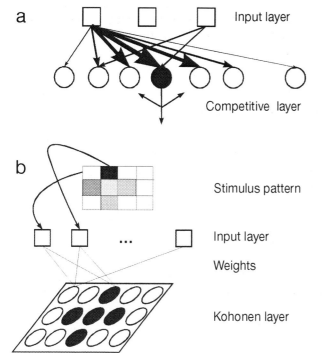

a   Input layer

Competitive layer

b   Stimulus pattern

Input layer

Weights

Kohonen layer

Fig. 8-13.
Structure of self-organizing networks in linear arrangement of the competitive layer (a) and in the Kohonen representation (b)

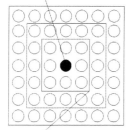

Winning neuron

1. Neighborhood

nen network. The pecularity of this network type is the maintenance of topology or, more generally, the pattern of the data vector to be learned.

The preservation of topology results from the introduction of neighborhood relationships between neurons in the learning algorithm. These relationships are described by a neighborhood function with the distance measure d as the independent variable. This function characterizes the distance to the winning neuron:

$$nf = nf(d) \tag{8-32}$$

Typical distance functions are based on triangular functions or the Mexican hat function (Fig. 8-14).

Fig. 8-14.
Neighborhood function in form of a triangle (A) and a Mexican hat (B)

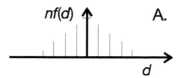

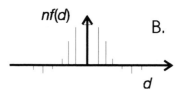

The learning algorithm for the Kohonen net operates by analogy with the competitive learning laws in Eqs. 8-19 to 8-21:
- Presentation of an input vector x.
- Calculation of the distance between the input vector and each output unit i to evaluate the winning neuron j:

$$\left\| w_j(t) - x(t) \right\| = \min_i \left\| w_i(t) - x(t) \right\| \tag{8-33}$$

- Change of the weight for the winning neuron and its neighbors by use of the neighborhood function is performed according to Eq. 8-32:

$$w_{ik}(t+1) = w_{ik}(t) + \eta_t \, nf(d)[x_i(t) - w_{ik}(t)] \tag{8-34}$$

where k is the neighborhood element at time t.

- Continuation of the learning process with subsequent x-pattern.

As a result of learning the patterns are arranged in clusters, presupposed the data vectors can be grouped in a Kohonen layer. The clusters can now be explored for assigning objects to them.

The result of unsupervised Kohonen learning, however, can also be used for *classification*. For this an additional layer is introduced. The output information trained by the Kohonen net is further trained on the known patterns or class information by means of an associative learning law. After adjustment of the additional weights the net can be subsequently applied for classifications.

Applications of neural networks are given in Table 8-4. The applications are, in principal, similar to those discussed in the statistics Chapters 5 and 6, i.e. grouping, classification and modeling of data. In addition, the nets also serve the purpose of knowledge processing, e.g., for machine learning of rules.

If used for modeling it must be noted that estimation of confidence intervals for the weights (parameters) or for predictions cannot be performed analytically because of the nonlinearity of the networks. As used with other nonlinear methods the confidence intervals must be estimated by means of Monte-Carlo simulations or bootstrapping methods.

At present investigations for applying neural networks in analytical chemistry are done without reference to statistical methods. Future studies will show which applications of the networks could be superior to existing multivariate methods including nonlinear statistical methods.

More advanced models are already available, e.g., those based on adaptive resonance theory (ART nets).

Bootstrapping is based on the use of the original data set to estimate confidence intervals. Part of the data (rows of the data matrix) is sorted out and used for later predictions. The missing rows in the matrix are replaced randomly by the retained data vectors. The latter vectors are then used twice in a computational run.

Table 8-4.
Applications of neural networks in analytical chemistry

Objective	Neural Net	Area	Reference
Parameter estimation	Backpropagation	Quantitative analysis in NIR	[8-14]
Clustering	Kohonen	IR spectra grouping	[8-15]
Classification	Backpropagation	MS structure elucidation	[8-16]
Identification	Backpropagation	HPLC diode array detection	[8-17]

8.3 Fuzzy theory

The theory of fuzzy sets enables representation and processing of vague propositions and uncertain information. In contrast with probability theory fuzzy-theory is a possibilistic approach, i.e. it is based on possibility theory. The concept of fuzzy sets was introduced by Lotfi A. Zadeh of the University of California (USA) in 1965.s

Fundamentals

Fuzzy sets

Venn diagram

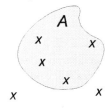

In classical set theory the containment of an element x to a subset A of the universe of discourse X is described by a characteristic function. It is called the membership function, $m(x)$. A membership value of 1 is assigned to an element x if it is contained in a set A. If x is not an element of the set A, a membership value of zero results:

$$m(x) = \begin{cases} 1 & \text{for} \quad x \in A \\ 0 & \text{for} \quad x \notin A \end{cases} \tag{8-35}$$

The concept of fuzzy sets extends this crisp assignment by allowing membership values *between* 0 and 1. These fuzzy sets are normalized to the interval [0,1]. They can be represented either discretely or by means of a function (cf. Eq. 8-36 to 8-39).

Example: 8-7: *Set theory*

The set of all elements detectable by an analytical method can be expressed by a set over the atomic number, x. According to classical set theory detection of elements with atomic numbers greater than a particular value is feasible, e.g. X-ray fluorescence analysis can be used for analyses of elements from sodium with atomic number 11. The membership function is given by the solid line in Fig. 8-15.

Fig. 8-15.
Membership function for the detectability of elements by X-ray fluorscence analysis. Solid line: classical (crisp) set; broken line line: fuzzy set

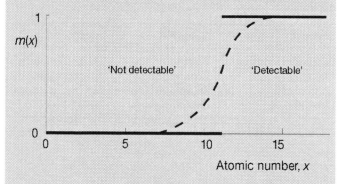

In practice, however, there is a transition between the detectability and non-detectability of an element. Thus it is not unusual with this analytical principle, given instrumentation of reasonable quality, that even the element with atomic number 9 (fluorine) can be detected. The transition between the set of detectable and non-detectable chemical elements is better described by a membership function based on a fuzzy set. This membership function is plotted in Fig. 8-15 as broken line.

Types of membership function

For characterization of a fuzzy set many possibilities exist. Depending on the objective, different types of membership function are used.

Fuzzy observations

Consider first the description of uncertainty of experimental observations, such as the uncertain position of a spectroscopic line.

The fuzzy set is characterized, e.g., by means of an exponential function of the type:

Note: membership functions are *not* probability distribution functions. A membership function can be specified for a *single* observation.

$$m(x) = \exp\left(-\frac{|x-a|^2}{b^2}\right) \tag{8-36}$$

Here a is equal to the x-value with membership value $m(x) = 1$. The constant $1/b^2$ normalizes the membership function to the interval $[0,1]$. As an example Fig. 8-16a depicts exponential membership functions with $a = 9$ and $b = 3$.

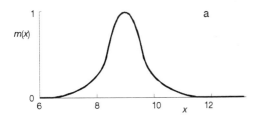

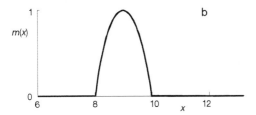

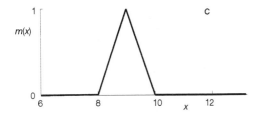

Fig. 8-16.
Membership functions for characterization of the uncertainty of experimental observations in the form of exponential (a), quadratic (b) and linear (c) functions

For operations with exponential functions it is recommended that they be truncated at a certain spread by setting the membership value, $m(x)$, to zero. Some types of membership function provide a more 'natural' truncation, e.g. a quadratic function of the type:

$$m(x) = \left[1 - \frac{|x-a|^2}{b^2} \right]^+$$

(8-37)

The plus sign denotes the restricted validity for positive values of the membership function. Negative values of $m(x)$ are set to zero (Fig. 8-16b).

Important is the *monotonicity* of the membership function. The special form of the membership function has only a weak influence on the result of fuzzy operations. The parabolic function in Eq. 8-37 can be approximated, therefore, by a triangular function (Fig. 8-16c):

$$m(x) = \left[1 - \frac{|x-a|}{b} \right]^+$$

(8-38)

+ again denotes truncation of the function to positive memberhip values.

In general, there are no restrictions on specification of membership functions. They can be based either on experimental observations or on experience. In addition, membership functions are not restricted to a single x-variable. A membership function for the dependence of experimental observations i on the variables x and y could, e.g., have the following form:

$$m_i(x, y) = \left\{ 1 - \left[\frac{|x-x_i|}{u_i} + \frac{|y-y_i|}{v_i} \right] \right\}$$

(8-39)

where u_i and v_i are constants.

This equation describes a pyramid-shaped membership function (cf. Fig. 8-17).

Linguistic variables

Verbal fuzzy expressions are described by fuzzy-theory by means of linguistic variables; for example the solubility of a substance in water can be characterized by the fuzzy terms 'high' and 'low'. If the linguistic variables are to be distinguished further, so-called *modifiers* are applied.

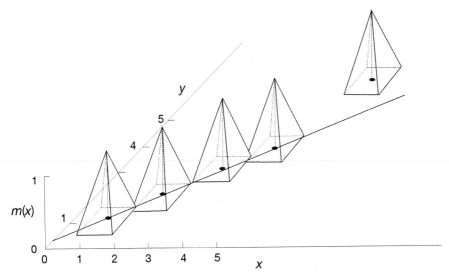

Fig. 8-17.
Two-dimensional membership function for fitting a straight line

Fig. 8-18 provides examples of linguistic modifiers, such as 'very', 'more or less' and 'middle'. The membership function '*high*' is normalized to the interval [0,1] and reads:

$$m_h(x) = \begin{cases} 0 & \text{for} \quad x < 0 \\ 2x^2 & \text{for} \quad 0 \leq x \leq 0.5 \\ 1 - 2(x-1)^2 & \text{for} \quad 0.5 \leq x \leq 1 \\ 1 & \text{for} \quad x > 1 \end{cases} \qquad (8\text{-}40)$$

Calculation of the modifiers is explained in Table 8-5.

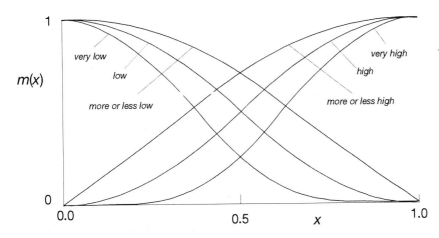

Fig. 8-18.
Membership functions for the linguistic variable 'high' according to Eq. 8-40 together with different modifiers

Table 8-5.

Calculation of modifiers for the membership functions of linguistic variables regarded as 'high' according to Eq. 8-40

Modifier	Formula
very high	$m_{vh}(x) = m(x)^2$
more or less high	$m_{mlh}(x) = \sqrt{m(x)}$
low (not high)	$m_n(x) = 1 - m(x)$
very low	$m_{vn}(x) = [1 - m(x)]^2$
more or less low	$m_{mln}(x) = \sqrt{1 - m(x)}$
middle	$m_m(x) = \min\{m(x), 1 - m(x)\}$

Truth values

For applications of multi-valued logic, in addition to the binary truth values 'true' and 'false' we need further graduation. For this we construct membership functions, that are normalized in the universe of discourse X to the interval [0,1]. The membership function is expressed, e.g., by the functions introduced by Baldwin (Fig. 8-19).

Fig. 8-19.
Membership functions for truth values after E. Baldwin

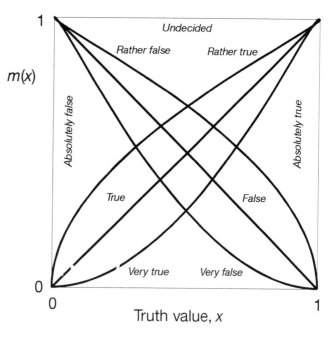

Discrete membership functions

Often specification of continuous membership functions does not make sense. If, for example, the qualification of a sample state for a spectrophotometric analysis is to be described, discrete values, such as solid, gas or liquid are required.

272

Discrete membership functions in relation to the fuzzy set A are represented as pairs consisting of the variable value and the corresponding membership value, e.g.:

$$A = |(1,0.5), (2,0.7)(3,1.0)| \qquad (8\text{-}41)$$

Operations with fuzzy sets

Fuzzy set operations are derived from classical set theory. In addition, there are theories for calculating with fuzzy numbers, functions, relations, measures or integrals.

The *intersection* of two sets A and B corresponds according to classical theory to all elements, that are simultaneously contained in both sets. For two fuzzy sets the intersection, $A \cap B$, is derived from the minimum of both of the membership functions $m_A(x)$ and $m_B(x)$:

$$m_{A \cap B}(x) = \min\{m_A(x), m_B(x)\} \qquad (8\text{-}42)$$

Intersection

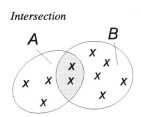

Two common sets are unified by aggregating into one set all elements that belong to at least one of the two sets. The *union* of fuzzy sets results from calculating the maximum of their membership functions:

$$m_{A \cup B}(x) = \max\{m_A(x), m_B(x)\} \qquad (8\text{-}43)$$

Union

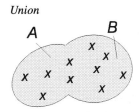

The *complement* of a set A corresponds to all elements, that are not contained in A. By analogy the complement of a fuzzy set A is:

$$m_{\bar{A}}(x) = 1 - m_A(x) \qquad (8\text{-}44)$$

The number of elements in a set is, in classical theory, termed *cardinality*. The latter is obtained for fuzzy sets from summation or integration over the membership function of that set:

$$cardA = \sum_i m(x_i) \quad \text{or} \quad cardA = \int_X m(x)\,dx \qquad (8\text{-}45)$$

Cardinalities in the interval $[0,1]$ are obtained by calculating their *relative cardinalities*. For this the standard set S is taken as the basis:

$$\text{rel}_S\,cardA = \frac{cardA}{cardS} \qquad (8\text{-}46)$$

Fig. 8-20 illustrates the set operations.

Example 8-8: *Cardinality of a fuzzy set*

The relative cardinality of the fuzzy set in Fig. 8-20c must be calculated. The membership function for this set is:

$$m(x) = \left[1 - |x-a|^2\right]^+ \tag{8-47}$$

Integration over the membership function in the inverval $x = 2$ to $x = 4$ provides for the relative cardinality according to Eq. 8-46:

$$\mathrm{rel}_s\, cardA = \frac{\int_2^4 \left[1 - |x-3|^2\right] dx}{4-2} = 0.666 \tag{8-48}$$

Fig. 8-20.
Intersection (a) and union (b) of two fuzzy sets, and the cardinality of a fuzzy set (c)

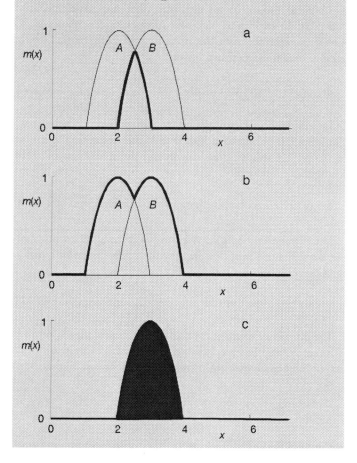

Applications

Fuzzy theory can be used either for data analysis or for dealing with fuzzy logic. Typical applications in analytical chemistry are [8-19]:

- Pattern recognition based on unsupervised and supervised learning
- Multi-criteria optimization
- Comparisons of spectra, chromatograms or depth profiles on the basis of fuzzy functions
- Fuzzy modeling
- Fuzzy logic for fuzzy control and for approximate reasoning

One example of the *grouping* of data on the basis of unsupervised learning has already been discussed in respect of fuzzy cluster analysis by the c-means algorithm (Sec. 5.2.2).

Fuzzy methods are useful for solving problems of *supervised learning*, if irregularly cast data sets are to be handled. For example, in the analysis of human cell tissues by means of capillary gas chromatography the number of chromatographic peaks (components) varies by ± 15 peaks per chromatogram. This is because of high biological variability and not random (analytical) error. A statistical procedure for pattern recognition cannot be used here, because the vectors (chromatograms) to be compared would require an equal number of elements in the vector, i.e. the same number of peaks in the chromatogram. By means of set theory approaches data vectors of different length can be evaluated.

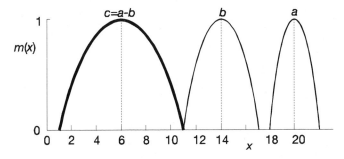

Fig. 8-21.
Example of a fuzzy difference between two numbers, i.e. approximately 20 minus approximately 14 gives approximately 6

For *multi-criteria optimization* the individual criteria are described by means of fuzzy sets and are then aggregated to an appropriate objective function. To define the membership functions of those objective functions heuristic knowledge can be included.

For fuzzy modeling or for comparison of fuzzy functions the fundamentals of *fuzzy arithmetic* are needed. These fundamentals are given, e.g., in references [8-18] or [8-19]. Applications are known for calibration of analytical methods and for qualitative and quantitative comparison of chromatograms, spectra or depth profiles.

Another important area of application is approximate reasoning based on *fuzzy logic*. Fuzzy inferences are applied in analytical expert systems or for controlling chemical or biotechnological reactors.

Table 8-6 lists some applications of fuzzy theory in analytical chemistry.

Table 8-6.
Applications of fuzzy theory in analytical chemistry [8-19]

Aim	Fuzzy method	Area
Unsupervised learning	fuzzy clustering	types of beer
Supervised learning	fuzzy pattern recognition	chromatograms
Library search/identification	set operations	IR spectra
Quality control	fuzzy functions	depth profiles, spectra
Modeling	fuzzy numbers	calibration with errors in x and y
Multi-criteria optimization	fuzzy sets	enzymatic determination
Expert system	fuzzy logic	X-ray fluorescence analysis [8-8]

8.4 Genetic algorithms

Natural computation denotes the following methods: simulated annealing, simulated evolution strategy, genetic algorithms, artificial immunonetworks, artificial neural nets, artificial chemical reaction systems and artificial life (cellular atoms and fractals).

Genetic algorithms belong to methods that have a biological analog as do neural nets. These methods are based on *evolutional components*, such as population of living beings, competition between living beings and reproduction of life. As a result new living beings are formed, that are better developed or those living beings that have reached the end of their lives must be replaced.

Genetic algorithms are especially useful for dealing with highly complex and highly dimensional *search problems*. This can be related to optimization of a set of variables in the sense of feature selection. A typical application is the selection of wavelength in spectrometric multicomponent analysis (cf. Sec. 6.2.3). A second application concerns optimization of parameters, e.g. for multi-criteria optimization or for parameter estimation in non-linear models. Here multi-peaked objective areas are operating. Many other optimization methods might not find the global optimum in such cases.

Apart from those numerical applications genetic algorithms can also be used to solve combinatorial problems, e.g. the processing of a sample queue in a laboratory.

Heredity in the computer

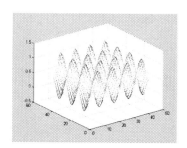

The starting point with genetic algorithms is a population of *living beings*. A computer-adaptable living being consists of a particular set of variables. Each variable is represented by a *chromosome* characterized by a set of genes. Usually the chromosomes are binary coded (cf. Sec. 1.3), so that each bit corresponds to a *gene*. If an 8-bit representation is used, then the chromosome consists of 8 genes and each gene can takes as an *allele* either the value 0 or 1.

Initial population

A genetic algorithm starts with an initial population of living beings, that are coded as character strings. Generation of the initial population can be performed, e.g., by random coin throws providing the bits 1 and 0 (cf. Example 8-8). Subsequent populations are generated by the genetic algorithm by *heredity* in the form of selection, crossover and mutation of the living beings.

8-Bit Chromosome

0	1	1	0	0	0	1	0

Selection

Selection or reproduction means multiplication as adaptation to environment or survival of dangers. Formaly the selection is based on the value of the objective function y for the living entity considered. In the simplest case the objective function is a signal. However, it can also be a composed quantity (cf. Sec. 4.1).

The new living being should develop a maximum (minimum) output signal. Typically the signal values are averaged. A good living being reveals a signal, that is greater than the average signal. A bad living being provides one that is lower than the average. The number of living beings prone to multiplication can be derived from the signal value that is weighted over the sum of all signals of the population.

Crossover

For the purpose of generation follows the crossover of living beings is exploited. Randomly a father and mother are selected. Also a crossover position is fixed in the chromosome or in the bit character string that is used to derive the aggregation of the genes of the parents. All bits left of the crossover point are transferred to the child from the father and all bits right of the point are given by the mother.

Mutation

Mutations are changes as a result of external effects. For individual bits it is decided on the basis of a predefined probability whether a bit switches or not.

A prerequisite for a well operating genetic algorithm is a well performing random generator. However, one should note that genetic algorithms are not based solely on chance (randomness), as is valid for Monte-Carlo simulations. Randomness here is only a tool for successfully proceeding in the space of the objective area.

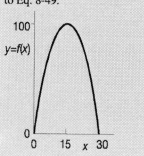

Objective function according to Eq. 8-49:

Example 8-9: *Genetic algorithm*

By means of a genetic algorithm the optimum dependence of a response (objective function) y on the variable x is sought. To facilitate the understanding and to be able to follow the calculations the response values are generated by the function in Eq. 8-49. In practice the repsonses are experimental observations.

$$y = \left[100 - 0.44(x - 15)^2\right]^+ \tag{8-49}$$

Initially variable x must be coded as a chromosome. For the x-values of interest here between 0 and 30 a single 5 bit-string is sufficient. Random selection of the initial population reveals the following four living beings (cf. x-values in Table 8-7):

```
1 0 0 1 1
0 0 0 0 1
1 1 0 1 0
0 0 1 0 1
```

For *selection* of successors the corresponding responses are measured or calculated according to Eq. 8-49. Then the living beings can be judged in relation to their adaptational ability. The new population must be formed from the surviving examples.

Table 8-7 summarizes the individual steps of the genetic algorithm. The four-character strings of the initial population are given with the corresponding x-values and the responses y in columns three and four. The weights of the y-values related to the sum of all responses of the population (selection probability) are higher than the average for living being 1, but the response for living being 2 is obviously below the population average. In the sense of propagation, more frequent reproduction is expected for living being 1 than for being 2. The expected number of reproductions can be estimated from the relationship with the average of the response values (column 6 of Table 8-7). As the discrete number of reproductions we obtain for living being 1 two copies, for being 2 no copy and for living beings 3 and 4 one copy each. The new population is then:

```
1. 1 0 0 1 1
2. 1 0 0 1 1
3. 1 1 0 1 0
4. 0 0 1 0 1
```

Table 8-7.

Selection in the genetic algorithm in Example 8-8

Nr.	Initial population	x	$y = f(x)$	$\dfrac{y_i}{\sum y}$	Expected number, $\dfrac{y_i}{\bar{y}}$	Actual number
1	1 0 0 1 1	19	93	0.44	1.77	2
2	0 0 0 0 1	1	14	0.07	0.27	0
3	1 1 0 1 0	26	47	0.22	0.89	1
4	0 0 1 0 1	5	56	0.27	1.07	1
	Sum:		210	1.00	4.00	4
	Average:		52.5	0.25	1.00	1

In the next step *crossover* of living beings is carried out. In our example living beings 1 and 3 and 2 and 4 are chosen randomly from the new population as pairs for crossover. Crossover points are the positions 4 and 2. As a result one obtains the crossovers given in Table 8-8.

Table 8-8.

Crossovers in the genetic algorithm in Example 8-8

Nr.	Father	Mother	Crossover point	Child	x	$y = f(x)$
1	1 0 0 1|1	1 1 0 1|0	4	1 0 0 1 0	18	96
2	1 0 0 1|1	1 1 0 1|0	4	1 1 0 1 1	21	84
3	1 0|0 1 1	0 0|1 0 1	2	1 0 1 0 1	27	37
4	1 0|0 1 1	0 0|1 0 1	2	0 0 0 1 1	3	37
	Sum:					254
	Average:					63.5

As a result of crossover a population emerges for which the sum of y-values or their average becomes larger (cf. Tables 8-7 and 8-8). So we have approached the maximum of the objective function more closely.

An additional change in the population would be feasible on the basis of mutations. For a probability of e.g. 0.005 in total 20 of the transferred bits should switch, i.e. $20 \times 0.005 = 0.1$. This means practically, that no bit will mutate from 0 to 1 or vice versa.

Applications

The applications of genetic algorithms are determined by some particularities which are:

- The algorithms do not operate on the basis of the input variables themselves but on coded variables;
- The search for the optimum starts from several points;
- The criterion for optimization is the objective function and not, e.g., its derivative;
- The forward movements are not deterministically based, but probabilistically.

Traditional search methods move in the objective area from one point to the next on the basis of a given search rule. In multimodal objective areas one would end up as a result of this point-by-point forward movement at the closest peak, which might be far from the global optimum. Genetic algorithms start on the basis of a population of points and move from those positions in a parallel fashion in the direction of the optimum.

Most frequently genetic algorithms are used for wavelength selection in multicomponent analysis. If, for example five wavelengths must be chosen out of 30 wavelength in total, then a 30-bit chromosome could be defined. The actual wavelength would be coded by ones and all other bits are set to zero. If the wavelength 3, 7, 13, 22 and 27 were tested, then the chromosome would look like this:

001000100000100000000100001000

Table 8-9.
Applications of genetic algorithms (see [8-22])

Aim	Strategy	Area
Learning algorithm for neural nets	optimization of weights	X-ray fluorescence analysis
Wavelength selection	selection of subsets	multicomponent analysis
Multi-criteria optimization	optimization	atomic emission spectrometry
Prediction of retention data	numerics	HPLC

Further applications range from multi-criteria optimization to the replacement of the backpropagation learning algorithm for training neural networks (Sec. 8.2) by means of genetic learning algorithms (Table 8-9).

8.5 General reading

Sec. 8.1 Artificial intelligence and expert systems

[8-1]   Cherniak, E. , McDermott, D., *Introduction to Artificial Intelligence*, Addison-Wesley, Reading, MA, 1985.
[8-2]   Winston, P. H., Horn, B. K. P., *Lisp*, 2. Edition, Addison-Wesley, Reading, MA, 1984.
[8-3]   Lindsay, R. K. et al., *Applications of Artificial Intelligence for Organic Chemistry – the Dendral project*, McGrawHill, New York, 1980.
[8-4]   Gunasingham, H., Srinivasan, B., Ananda, A. L., Anal. Chim. Acta **182** (1986) 193.
[8-5]   Buydens, L. M. C., Schoenmakers, P. J. (Eds.), *Intelligent software for chemical analysis*, Elsevier, Amsterdam, 1993.

[8-6] Peichang, L., Hongxin, H., J. Chromatography **452** (1988) 175.

[8-7] Janssens, K., Van Espen, P., Anal. Chim. Acta **184** (1986) 117.

[8-8] Arnold, T., Otto, M., Wegscheider, W., Talanta, **41** (1994) 1169–1184.

[8-9] Wünsch, G., Gansen, M., Anal. Chim. Acta **333** (1989) 607.

[8-10] Wagner, A., Flock, J., Otto, M., Stahl und Eisen 115 (1995) 57–60.

Sec. 8.2 Neural networks

[8-11] Rumelhart, D. E., McClelland, J. L., *Parallel Distributed Processing, Explorations in the Microstructure of Cognition*, Vol. 1: Foundations, Vol. 2: Psychological and Biological Models, The MIT Press, Cabridge, MA, 1986.

[8-12] Kohonen, T., *Self-Organization and Associative Memory*, Springer, Berlin, 1984.

[8-13] Kosko, B., *Neural Networks and Fuzzy Systems*, Prentice-Hall, London, 1992.

[8-14] Borggaard, C., Thodberg, H. H., Anal. Chem. **64** (1992) 545.

[8-15] Zupan, J., Gasteiger, J., *Neural Networks for Chemists*, VCH, Weinheim, 1993.

[8-16] Curry, B., Rumelhart, D.E., Tetrahedron Comput. Methodol. **3** (1990) 213-237.

[8-17] Mittermayr, C. R., Drouen, A. C. J. H., Otto, M., Grasserbauer, M., Anal. Chim. Acta **294** (1994) 227–242.

Sec. 8.3 Fuzzy Theory

[8-18] Bandemer, H., Näther, W., *Fuzzy Data Analysis*, Theory and Decision Library, Series B: Mathematical and Statistical Methods, Kluwer Academic Publishers, Dordrecht, 1992.

[8-19] Otto, M., *Fuzzy Theory Explained*, Chemometrics Intell. Lab. Syst. **1** (1986) 71.

[8-20] Bezdek, J. C., *Pattern Recognition with Fuzzy Objective Function Algorithms*, Plenum Press, New York, 1982.

Sec. 8.4 Genetic algorithms

[8-21] Goldberg, D. E., Genetic *Algorithms in Search, Optimization, and Machine Learning*, Addison-Wesley, Reading, MA, 1989.

[8-22] Lucasius, C. B., Kateman, G., Chemometrics Intell. Lab. Syst. **19** (1993) 1.

[8-23] Kalivas, J. H. (Ed.), *Adaption of Simulated Annealing to Chemical Optimization Problems*, Elsevier, Amsterdam, 1995.

Questions and problems

1. What are the differences between PROLOG, LISP and an expert system shell?
2. Discuss the difference between search by use of the simplex method and by use of an in-depth strategy.
3. Summarize typical signal transfer functions used with artificial neural networks.
4. How would you arrange the neurons in a multi-layer perceptron to classify gas chromatograms consisting of 20 peaks and belonging to four different classes?
5. What is the difference between associative and competitive learning laws?
6. Which neural network preserves the topological structure of data?
7. What is the difference between probability and possibility theory and which methods represent the two particular approaches?
8. How are genetic algorithms used for feature selection in the framework of pattern recognition?

9 Quality assurance and good laboratory practice

Learning objectives

- To introduce the tools available to ensure the quality of analytical-chemical measurement
- To learn about regulatory and legal aspects of quality assurance and quality control

Analytical data need to be comparable between different laboratories on an international scale. This supposes that the quality of the data is assured. In addition, planning, performance, reporting and archiving of tests should be regulated.

A central point of comparability of analytical data is an appropriate system of quality assurance. According to DIN 55350 *quality* is defined by:

The total of properties and features of a product
or an operation to fulfil predefined requirements.

To quantify quality in the framework of quality assurance, tests become necessary. *Quality assurance* comprises all activities that lead to obtention of the defined requirements. They include the totality of operations in quality management, quality planning, quality directing and quality tests.

If the requirement cannot be fulfilled then an *error* is made according to the DIN directive. If the error restricts the applicability it is termed a *lack*. Typical of analyses are random and systematic errors, false positive or false negative detection results or the complete failure of an analytical determination.

To control errors in an analysis, the individual steps of the procedure must be fixed in detail. In addition, the analytical procedure must be tested for its performance, i.e. it must be *validated*. For the use of procedures in routine analysis additional checks are necessary that will be explained below.

9.1 Validation and quality control

Validation in analytical chemistry

Validation describes, in general, the assurance that an analytical procedure provides reproducible and secure results that are suitable for the application intended.

Validation criteria are:
Trueness
Precision
Dynamic range
Selectivity
Limit of detection
Limit of determination
Robustness

The first step in developing a relative analytical method is *calibration*. It is based on the use of standard solutions or of solid standards. The calibration function (Eq. 4-1) is constructed by means of linear regression analysis as discussed in Sec. 6.1.

The *detection limit* is that analyte concentration, that corresponds to the averaged signal of the blank plus three times the standard deviation.

The precision of the procedure can be characterized by the *standard deviation of the procedure*. As additional performance characterisics the *limit of detection* (Eq. 4-3) and *working range* are reported.

To test for systematic deviations, caused by the influences of different steps of an analytical procedure or by the sample matrix, the recovery (rate) is calculated (cf. Eq. 4-6). To investigate an analytical procedure for systematic devations, the recovery function is applied.

Recovery function

The *limit of determination* is the lowest analyte concentration, that can be determined with acceptable accuracy.

The recovery function describes the relationship between the found, x_{found}, and true, x_{true}, concentration by means of a straight-line model:

$$x_{found} = a_0 + a_1 x_{true} \qquad (9\text{-}1)$$

The *working range* of an analytical method denotes the range between the lower and upper concentrations for which accurate determinations are feasible.

where a_0 and a_1 are the regression parameters. In an ideal case the recovery function should pass through the origin of the coordinate system and have a slope of 1, i.e.:

$$a_0 = 0 \text{ and } a_1 = 1.0 \qquad (9\text{-}2)$$

In practice these conditions will be valid only approximately. The test for the significance of the deviations can be carried out by means of the confidence intervals for the parameters a_0 and a_1 on the basis of a *t*-test (cf. Eq. 6-47). The confidence intervals for the parameters are:

Accuracy stands for both trueness and precision.

$$\Delta a_0 = a_0 \pm t(1 - \alpha/2; f)s_{a_0} \qquad (9\text{-}3)$$

$$\Delta a_1 = a_1 \pm t(1 - \alpha/2; f)s_{a_1} \qquad (9\text{-}4)$$

where t is the quantile of the Student distribution (Table III), α is the significance level, and f is the number of degrees of freedom.

The standard deviations s_{a_0} and s_{a_1} for the parameters are calculated according to Eqs. 6-7 and 6-8 in Sec. 6.3. A constant systematic deviation is given at a significance level α if the confidence interval Δa_0 does not include the value $a_0 = 0$. When the confidence interval Δa_1 does not contain the value $a_1 = 1$ a proportional systematic deviation is valid.

Finally the *robustness* of analytical procedures must be investigated. This has to ensure that the quality of data is independent of small variations during the performance of the procedure.

Robustness can be checked by laboratory intercomparison studies, the performance of which will be discussed in the section on external quality control. In ones own laboratory robustness can be checked by variation of the experimental variables within tolerable limits.

The elaborated analytical procedure is the basis of quality assurance in routine analysis. It is summarized in a *standard operating procedure (SOP)*. The SOP defines the range of application and the goals of quality.

The standard deviation of a procedure is calculated by use of Eq. 6-9:

$$s_y = \sqrt{\frac{\sum_{i=1}^{n}(y_i - \hat{y}_i)^2}{n-2}}$$

On the basis of the concentration, x, we obtain:

$$s_x = \frac{s_y}{b_1}$$

Internal quality assurance

Quality control of a procedure in routine analysis is based primarily on evaluation of quantities that characterize precision and reliability, such as mean, standard deviation, dynamic range, recovery rate and reliability ranges (dispersion and confidence interval).

For internal quality assurance *control samples* are applied. These include:

- Standard solutions
- Blank samples
- Real samples
- Synthetic samples
- Certified standard reference materials

The control samples should be analysed at least once or twice in each series of analyses, to monitor the accuracy of measurements.

Control charts

To monitor test procedures or to ensure the quality of products and processes by using analytical methods quality control charts have proven their success. There a quality characteristic is recorded at a given spacing in a chart enabling ready recognition of typical situations (Fig. 9-1).

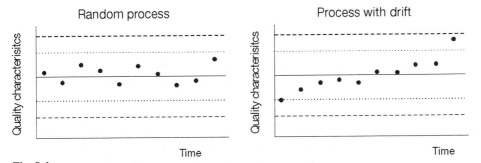

Fig. 9-1.
Sequence of quality characteristics in a control chart. The dotted line characterizes the lower and upper warning limits. The broken line describes the lower and upper control limits.

As quality characteristics the reference value and the control limits are used. We distinguish according to the kind of the quality characteristic individual value charts, mean charts, such as the $\bar{x}$-chart, median or blank chart, dispersion charts, such as standard deviation and range chart, and the recovery rate chart. Calculation of the quality characteristic plotted as central line and its thresholds is given in Table 9-1 for important control charts. The limits for the $\bar{x}$-chart are evaluated on the basis of a t distribution. In a standard deviation chart the thresholds are defined by applying the χ^2-distribution. The χ^2-quantiles can be taken from Table V in the appendix. The range charts are based on the ranges between the largest and smallest observations within a subgroup i ($x_{i,max}$ and $x_{i,min}$). The upper and lower limits result from multiplication with the D-factors given in Table 9-2 for 95% and 99% statistical certainty.

Table 9-1.
Control quantities in quality control charts. N – number of subgroups; n – number of replicate measurements per subgroup (n_i); x_i – observation; s_i – standard deviation from n_i parallel determinations; t – Student factor; P – probability; f –degrees of freedom.

Objective criterion	Calculation of objective criterion	Lower limit	Upper limit
Mean, $\bar{x}$	$\bar{x} = \dfrac{1}{n}\sum\limits_{i=1}^{n} x_i$	$\bar{x} - t(P,f)\dfrac{s}{\sqrt{n}}$	$\bar{x} + t(P,f)\dfrac{s}{\sqrt{n}}$
Mean standard deviation, s_m	$s_m = \sqrt{\dfrac{\sum\limits_{i=1}^{N}(n_i-1)s_i^2}{\sum\limits_{i=1}^{N}(n_i-1)}}$	$s_m\sqrt{\dfrac{1}{n-1}\chi^2\left(n-1;\dfrac{\alpha}{2}\right)}$	$s_m\sqrt{\dfrac{1}{n-1}\chi^2\left(n-1;1-\dfrac{\alpha}{2}\right)}$
Range, $\bar{R}$	$\bar{R} = \dfrac{\sum\limits_{i=1}^{N}(x_{i,max}-x_{i,min})}{N}$	$D_{lower}\bar{R}$	$D_{upper}\bar{R}$

The significance level of 5% serves as warning limit. A single crossing of the warning limit only requires increased attention to the control of the process. The control limit is fixed at a significance limit of 1%. If a value exceeds the control limit immediate action will be necessary. One denotes this as the outer-control-situation.

Description of a quality assurance system is provided in a *quality assurance handbook*. There all structures, responsibilities, standard operational procedures and tools for realization of the quality assurance are summarized.

Table 9-2.

D-factors for calculation of the limits of the range chart for probabilities *P* of 95% and 99%

	P = 95% or α = 5%		P = 99% or α = 1%	
n	D_{lower}	D_{upper}	D_{lower}	D_{upper}
2	0.039	2.809	0.008	3.518
3	0.179	2.176	0.080	3.518
4	0.289	1.935	0.166	2.614
5	0.365	1.804	0.239	2.280
6	0.421	1.721	0.296	2.100
7	0.462	1.662	0.341	1.986
8	0.495	1.617	0.378	1.906
9	0.522	1.583	0.408	1.846
10	0.544	1.555	0.434	1.798

External quality assurance

Laboratory intercomparison studies

These studies are performed to assure the comparability of analytical results. The aims are:

- Standardization of analytical procedures;
- Controling the analyses of a laboratory;
- Preparation of certified reference material.

According to DIN 38402 at least eight laboratories should participate in a study or still better more than 15 laboratories. These laboratories perform typically four parallel determinations each. Precision of the individual labs can be evaluated from parallel determinations on the basis of the standard deviations. To judge the trueness the recovery is determined according to Eq. 4-6. The laboratory means are related then to a true value or to the total mean.

Traceability

The results of chemical analyses should be comparable among each other as much as is feasible with physical quantities: the length of an object is given exactly in meters or the mass in kilograms.

The basis of comparability of chemical analyses is the relationship with standard reference materials, the contents or concentrations of which are known exactly. The elements and compounds contained in a sample are preferably traced back to the mol. Problems in traceability of chemical analyses frequently results from the limited selectivity of analytical methods. Determination of the mass of a pure substance can be done in principle without fundamental errors, but this is not a priori true for determination of the concentration of a single substance in a complex matrix such as blood. Validation of analytical results is therefore often more difficult than evaluation of physical quantities.

9.2 Accreditation and good laboratory practice

Accreditation of laboratories

An analytical laboratory proves its effective quality assurance system by *accreditation*. This is to attest to the competence of a laboratory to perform given analytical methods compared with independent assessment. In different countries different accreditation systems have emerged. For example, in Germany there is no central accreditation system, but a system that is decentralized to different sectors.

The Commission of the European Union has harmonized national accreditation systems. Unique criteria were developed for the operation of the test laboratories, and for their accreditation and certification. The result of those harmonization policies is the series of directives Euronorm EN 45000 (Table 9-3). All accreditations are performed at present on the basis of EN 45000.

Table 9-3.
General criteria in the directive series EN 45000

Euronorm	Topic
EN 45001	Running test laboratories
EN 45002	Evaluation of test laboratories
EN 45003	Organization of accreditation offices
EN 45011	Certification of products
EN 45012	Certification of personnel

Basics of good laboratory practice

What is understood by the system of good laboratory practice (GLP) which has existed for more than 20 years? GLP can be traced back to irregularities detected at the beginning of the 70s by the American Food and Drug Administration (FDA) during a review of toxicological studies. The results were regulations for good laboratory practice for testing of toxicology. They had to be complied with not only by companies in the USA, but by all countries exporting to the USA. The Organisation for Ecomomic Cooperation and Development (OECD) took over the internationalization of those standards. In Germany GLP is defined by the new law on chemicals.

GLP is mainly concerned with the assured repeatability of investigations. The assurance of quality is guaranteed in the system by an additional quality assurance unit monitored by continual inspections to maintain the principles of GLP.

The workload caused by the formalism and archiving is much larger than that of the EN 45000 directives. The GLP system is applied in areas where it is administered by law, e.g., in the case of the introduction of new chemicals, in toxicology or in health systems.

9.3 General reading

[9-1] Mayer, E. A., Griepink, D. B., Quality assurance, in Kellner, H., Mermet, J.-M., Otto, M., Widmer, M. (Eds.), *Analytical Chemistry*, Wiley-VCH, Weinheim, 1998.

[9-2] Günzler, H., (Ed.), Accreditation and quality assurance in analytical chemistry, Springer, Berlin, 1996.

Questions and problems

1. Define the analytical performance characteristics precision, trueness, accuracy, selectivity, dynamic range, working range, recovery, robustness, detection limit, and limit of determination.
2. What are reference materials used for?
3. What are control charts used for and which measures are common?
4. What is the meaning of 'traceability to the mol'?
5. What quality assurance system is applied in GLP?

Appendix

Statistical distributions

Table I.
Probability density function (ordinate values) of the standardized normal distribution

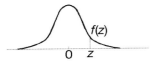

z	0.00	0.01	0.02	0.03	0.04	0.05	0.06	0.07	0.08	0.09
0.0	0.3989	0.3989	0.3989	0.3988	0.3986	0.3984	0.3982	0.3980	0.3977	0.3973
0.1	0.3970	0.3965	0.3961	0.3956	0.3951	0.3945	0.3939	0.3932	0.3925	0.3918
0.2	0.3910	0.3902	0.3894	0.3885	0.3876	0.3867	0.3857	0.3847	0.3836	0.3825
0.3	0.3814	0.3802	0.3790	0.3778	0.3765	0.3752	0.3739	0.3726	0.3712	0.3697
0.4	0.3683	0.3668	0.3653	0.3637	0.3621	0.3605	0.3589	0.3572	0.3555	0.3538
0.5	0.3521	0.3503	0.3485	0.3467	0.3448	0.3429	0.3411	0.3391	0.3372	0.3352
0.6	0.3332	0.3312	0.3292	0.3271	0.3251	0.3230	0.3209	0.3187	0.3166	0.3144
0.7	0.3132	0.3101	0.3079	0.3056	0.3034	0.3011	0.2989	0.2966	0.2943	0.2920
0.8	0.2897	0.2874	0.2850	0.2827	0.2803	0.2779	0.2756	0.2732	0.2709	0.2685
0.9	0.2661	0.2637	0.2613	0.2589	0.2565	0.2541	0.2516	0.2492	0.2468	0.2444
1.0	0.2420	0.2396	0.2371	0.2347	0.2323	0.2299	0.2275	0.2251	0.2227	0.2203
1.1	0.2179	0.2155	0.2131	0.2107	0.2083	0.2059	0.2036	0.2012	0.1989	0.1965
1.2	0.1942	0.1919	0.1895	0.1872	0.1849	0.1827	0.1804	0.1781	0.1759	0.1736
1.3	0.1714	0.1691	0.1669	0.1647	0.1626	0.1604	0.1582	0.1561	0.1539	0.1518
1.4	0.1497	0.1476	0.1456	0.1435	0.1415	0.1394	0.1374	0.1354	0.1334	0.1315
1.5	0.1295	0.1276	0.1257	0.1238	0.1219	0.1205	0.1182	0.1163	0.1445	0.1127
1.6	0.1109	0.1092	0.1074	0.1057	0.1040	0.1223	0.1006	0.09892	0.08728	0.09566
1.7	0.09405	0.09246	0.09089	0.08933	0.08780	0.08628	0.08478	0.08329	0.08183	0.08038
1.8	0.07895	0.07754	0.07614	0.07477	0.07341	0.0721	0.07074	0.06943	0.06814	0.06687
1.9	0.06562	0.06438	0.06316	0.06195	0.06077	0.0596	0.05844	0.05730	0.05618	0.05508
2.0	0.05399	0.05292	0.05186	0.05082	0.04980	0.0488	0.04780	0.04632	0.04586	0.04491
2.1	0.04398	0.04307	0.04217	0.04128	0.04041	0.03955	0.03871	0.03788	0.03706	0.03626
2.2	0.03547	0.03470	0.03394	0.03319	0.03246	0.03174	01.03103	0.03034	0.02965	0.02898
2.3	0.02833	0.02768	0.02705	0.02643	0.02582	0.02522	0.02463	0.02406	0.02349	0.02294
2.4	0.02239	0.02186	0.02134	0.02083	0.02033	0.01984	0.01936	0.01889	0.01842	0.01797
2.5	0.01753	0.01709	0.01667	0.01625	0.01585	0.01545	0.01506	0.01468	0.01431	0.01394
2.6	0.01358	0.01323	0.01289	0.01256	0.01223	0.01191	0.01160	0.01130	0.01100	0.01071
2.7	0.01042	0.1014	0.009871	0.009606	0.009347	0.00909	0.00885	0.00861	0.00837	0.00814
2.8	0.00792	0.00770	0.007483	0.007274	0.007071	0.00687	0.00668	0.00649	0.00631	0.00613
2.9	0.00595	0.00578	0.005616	0.005454	0.005296	0.00514	0.00499	0.00485	0.00471	0.00457
3.0	0.00443									

Example: $f(0.42) = 0.3653$

Table II.

Areas for the standard normal variate z (Eq. 2-28) of the normal distribution

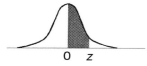

z	0.00	0.01	0.02	0.03	0.04	0.05	0.06	0.07	0.08	0.09
0.0	0.0000	0.0040	0.080	0.0120	0.0160	0.0199	0.0239	0.0279	0.0319	0.0359
0.1	0.0398	0.0438	0.0478	0.0517	0.0557	0.0596	0.0636	0.0675	0.0714	0.0753
0.2	0.0793	0.0832	0.0871	0.0910	0.0948	0.0987	0.1026	0.1064	0.1103	0.1141
0.3	0.1179	0.1217	0.1255	0.1293	0.1331	0.1368	0.1406	0.1443	0.1480	0.1517
0.4	0.1554	0.1591	0.1628	0.1664	0.1700	0.1736	0.1772	0.1808	0.1844	0.1879
0.5	0.1915	0.1950	0.1985	0.2019	0.2054	0.2088	0.2123	0.2157	0.2190	0.2224
0.6	0.2257	0.2291	0.2324	0.2357	0.2389	0.2422	0.2454	0.2486	0.2518	0.2549
0.7	0.2580	0.2612	0.2642	0.2673	0.2704	0.2734	0.2764	0.2794	0.2823	0.2852
0.8	0.2881	0.2910	0.2939	0.2967	0.2995	0.3023	0.3051	0.3078	0.3106	0.3133
0.9	0.3159	0.3186	0.3212	0.3238	0.3264	0.3289	0.3315	0.3340	0.3365	0.3389
1.0	0.3413	0.3438	0.3461	0.3485	0.3508	0.3531	0.3554	0.3577	0.3599	0.3621
1.1	0.3643	0.3665	0.3686	0.3708	0.3729	0.3749	0.3770	0.3790	0.3810	0.3830
1.2	0.3849	0.3869	0.3888	0.3907	0.3925	0.3944	0.3962	0.3980	0.3997	0.4015
1.3	0.4032	0.4049	0.4066	0.4082	0.4099	0.4115	0.4131	0.4147	0.4162	0.4177
1.4	0.4192	0.4207	0.4222	0.4236	0.4251	0.4265	0.4279	0.4292	0.4306	0.4319
1.5	0.4332	0.4345	0.4357	0.4370	0.4382	0.4394	0.4406	0.4418	0.4429	0.4441
1.6	0.4452	0.4463	0.4474	0.4484	0.4495	0.4505	0.4515	0.4525	0.4535	0.4545
1.7	0.4554	0.4564	0.4573	0.4582	0.4591	0.4599	0.4608	0.4616	0.4625	0.4633
1.8	0.4641	0.4649	0.4656	0.4664	0.4671	0.4678	0.4686	0.4693	0.4699	0.4706
1.9	0.4713	0.4719	0.4726	0.4732	0.4738	0.4744	0.4750	0.4756	0.4761	0.4767
2.0	0.4772	0.4778	0.4783	0.4788	0.4793	0.4798	0.4803	0.4808	0.4812	0.4817
2.1	0.4821	0.4826	0.4830	0.4834	0.4838	0.4842	0.4846	0.4850	0.4854	0.4857
2.2	0.4861	0.4864	0.4868	0.4871	0.4875	0.4878	0.4881	0.4884	0.4887	0.4890
2.3	0.4893	0.4896	0.4898	0.4901	0.4904	0.4906	0.4909	0.4911	0.4913	0.4916
2.4	0.4918	0.4920	0.4922	0.4925	0.4927	0.4929	0.4931	0.4932	0.4934	0.4936
2.5	0.4938	0.4940	0.4941	0.4943	0.4945	0.4946	0.4948	0.4949	0.4951	0.4952
2.6	0.4953	0.4955	0.4956	0.4957	0.4959	0.4960	0.4961	0.4962	0.4963	0.4964
2.7	0.4965	0.4966	0.4967	0.4968	0.4969	0.4970	0.4971	0.4972	0.4973	0.4974
2.8	0.4974	0.4975	0.4976	0.4977	0.4977	0.4978	0.4979	0.4979	0.4980	0.4981
2.9	0.4981	0.4982	0.4982	0.4983	0.4984	0.4984	0.4985	0.4985	0.4986	0.4986
3.0	0.49865	0.4987	0.4987	0.4988	0.4988	0.4989	0.4989	0.4989	0.4990	0.4990
4.0	0.49997									

Table III.

Two- and one-sided Student's *t* **distributions** for different risk levels α and degrees of freedom from $f = 1$ to $f = 20$

Two-sided Student's *t*-distribution: One-sided Student's *t*-distribution:

f	$\alpha = 0.05$	$\alpha = 0.01$	f	$\alpha = 0.05$	$\alpha = 0.025$
1	12.706	63.657	1	6.314	12.706
2	4.303	9.925	2	2.920	4.303
3	3.182	5.841	3	2.353	3.182
4	2.776	4.604	4	2.132	2.776
5	2.571	4.032	5	2.015	2.574
6	2.447	3.707	6	1.943	2.447
7	2.365	3.499	7	1.895	2.365
8	2.306	3.355	8	1.860	2.306
9	2.262	3.250	9	1.833	2.262
10	2.228	3.169	10	1.812	2.228
11	2.201	3.106	11	1.796	2.201
12	2.179	3.055	12	1.782	2.179
13	2.160	3.012	13	1.771	2.160
14	2.145	2.977	14	1.761	2.145
15	2.131	2.947	15	1.753	2.131
16	2.120	2.921	16	1.746	2.120
17	2.110	2.898	17	1.740	2.110
18	2.101	2.878	18	1.734	2.101
19	2.093	2.861	19	1.729	2.093
20	2.086	2.845	20	1.725	2.086

Table IV.

F-distribution for risk levels $\alpha = 0.025$ (lightface type) and $\alpha = 0.01$ (boldface type) for degrees of freedom f_1 and f_2

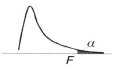

f_2	$f_1=1$	2	3	4	5	6	7	8	9	10	12	20	50	∞
1	647	779	864	899	922	937	948	956	963	968	976	993	1008	1018
	4052	**4999**	**5403**	**5625**	**5764**	**5859**	**5928**	**5981**	**6022**	**6056**	**6106**	**6208**	**6302**	**6366**
2	38.51	39.00	39.17	39.25	39.30	39.33	39.36	39.37	39.39	39.40	39.41	39.54	39.48	39.50
	98.49	**99.00**	**99.17**	**99.25**	**99.30**	**99.33**	**99.36**	**99.37**	**99.39**	**99.40**	**99.42**	**99.45**	**99.48**	**99.50**
3	17.44	16.04	15.44	15.10	14.88	14.73	14.62	14.54	14.47	14.42	14.34	14.17	14.01	13.90
	34.12	**30.82**	**29.46**	**28.71**	**28.24**	**27.91**	**27.67**	**27.49**	**27.34**	**27.23**	**27.05**	**26.65**	**26.35**	**26.12**
4	12.22	10.65	9.98	9.60	9.36	9.20	9.07	8.98	8.90	8.84	8.75	8.56	8.38	8.26
	21.20	**18.00**	**16.69**	**15.98**	**15.52**	**15.21**	**14.98**	**14.80**	**14.66**	**14.54**	**14.37**	**14.02**	**13.69**	**13.46**
5	10.01	8.43	7.76	7.39	7.15	6.98	6.85	6.76	6.68	6.62	6.52	6.33	6.14	6.02
	16.26	**13.27**	**12.06**	**11.39**	**10.97**	**10.67**	**10.45**	**10.29**	**10.15**	**10.05**	**9.89**	**9.55**	**9.24**	**9.02**
6	8.81	7.26	6.60	6.23	5.99	5.82	5.70	5.60	5.52	4.06	5.46	5.37	4.98	4.85
	13.74	**10.92**	**9.78**	**9.15**	**8.75**	**8.47**	**8.26**	**8.10**	**7.98**	**7.87**	**7.72**	**7.40**	**7.10**	**6.88**
7	8.07	6.54	5.89	5.52	5.29	5.12	4.99	4.90	4.82	4.76	4.67	4.47	4.28	4.14
	12.25	**9.55**	**8.45**	**7.85**	**7.46**	**7.19**	**7.00**	**6.84**	**6.71**	**6.62**	**6.47**	**6.16**	**5.87**	**5.65**
8	7.57	6.06	5.42	5.05	4.82	4.65	4.53	4.43	4.36	4.29	4.20	4.00	3.81	3.67
	11.26	**8.65**	**7.59**	**7.01**	**6.63**	**6.37**	**6.19**	**6.03**	**5.91**	**5.82**	**5.67**	**5.36**	**5.08**	**4.86**
9	7.21	5.71	5.08	4.72	4.84	4.32	4.20	4.10	4.03	3.96	3.87	3.67	3.47	3.33
	10.56	**8.02**	**6.99**	**6.42**	**6.06**	**5.80**	**5.62**	**5.47**	**5.35**	**5.26**	**5.11**	**4.81**	**4.53**	**4.31**
10	6.94	5.46	4.83	4.47	4.24	4.07	3.95	3.85	3.78	3.72	3.62	3.42	3.22	3.08
	10.04	**7.56**	**6.55**	**5.99**	**5.64**	**5.39**	**5.21**	**5.06**	**4.95**	**4.85**	**4.71**	**4.41**	**4.13**	**3.91**
12	6.55	5.10	4.47	4.12	3.89	3.73	3.61	3.51	3.44	3.37	3.28	3.07	2.87	2.72
	9.33	**6.93**	**5.95**	**5.41**	**5.06**	**4.82**	**4.65**	**4.50**	**4.39**	**4.30**	**4.16**	**3.86**	**3.58**	**3.36**
20	5.87	4.46	3.86	3.51	3.29	3.13	3.01	2.91	2.84	2.77	2.68	2.46	2.25	1.84
	8.10	**5.85**	**4.94**	**4.43**	**4.10**	**3.87**	**3.70**	**3.56**	**3.46**	**3.37**	**3.23**	**2.94**	**2.65**	**2.42**
50	5.34	3.98	3.39	3.05	2.83	2.67	2.55	2.46	2.38	2.32	2.22	1.99	1.75	1.56
	7.20	**5.08**	**4.22**	**3.74**	**3.45**	**3.21**	**3.04**	**2.91**	**2.81**	**2.72**	**2.58**	**2.29**	**1.98**	**1.70**
∞	5.02	3.69	3.12	2.79	2.57	2.41	2.29	2.19	2.11	2.05	1.94	1.71	1.43	1.00
	6.63	**4.61**	**3.78**	**3.32**	**3.02**	**2.80**	**2.64**	**2.51**	**2.41**	**2.32**	**2.18**	**1.88**	**1.53**	**1.00**

Table V.

Chi-square distribution for different degrees of freedom f at different probabilities P, $\chi^2(P;f)$

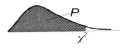

f	$P = 0.01$	0.05	0.10	0.25	0.50	0.75	0.90	0.95	0.99
1	0.00016	0.0039	0.0158	0.102	0.455	1.32	2.71	3.84	6.63
2	0.0201	0.103	0.211	0.575	1.39	2.77	4.61	5.99	9.21
3	0.115	0.352	0.584	1.21	2.37	4.11	6.25	7.81	11.3
4	0.297	0.711	1.06	1.92	3.36	5.39	7.78	9.49	13.3
5	0.554	1.15	1.61	2.67	4.35	6.63	9.24	11.1	15.1
6	0.872	1.64	2.20	3.45	5.35	7.84	10.6	12.6	16.8
7	1.24	2.17	2.83	4.25	6.35	9.04	12.0	14.1	18.5
8	1.65	2.73	3.49	5.07	7.34	10.2	13.4	15.5	20.1
9	2.09	3.33	4.17	5.90	8.34	11.4	14.7	16.9	21.7
10	2.56	3.94	4.87	6.74	9.34	12.5	16.0	18.3	23.2
11	3.05	4.57	5.58	7.58	10.3	13.7	17.3	19.7	24.7
12	3.57	5.23	6.30	8.44	11.3	14.8	18.5	21.0	26.2
13	4.11	5.89	7.04	9.30	12.3	16.0	19.8	22.4	27.7
14	4.66	6.57	7.79	10.2	13.3	17.1	21.1	23.7	29.1
15	5.23	7.26	8.55	11.0	14.3	18.2	22.3	25.0	30.6
16	5.81	7.95	9.31	11.9	15.3	19.4	23.5	26.3	32.0
17	6.41	8.67	10.1	12.8	16.3	20.5	24.8	27.6	33.4
18	7.01	9.39	10.9	13.7	17.3	21.6	26.0	28.9	34.8
19	7.63	10.1	11.7	14.6	18.3	22.7	27.2	30.1	36.2
20	8.26	10.9	12.4	15.5	19.3	23.8	28.4	31.4	37.6
21	8.90	11.6	13.2	16.3	20.3	24.9	29.6	32.7	38.9
22	9.54	12.3	14.0	17.2	21.3	26.0	30.8	34.0	40.3
23	10.2	13.1	14.8	18.1	22.3	27.1	32.0	35.2	41.6
24	10.9	13.8	15.7	19.0	23.3	28.2	33.2	36.4	43.0
25	11.5	14.6	16.5	19.9	24.3	29.3	34.4	37.7	44.3

Table VI.

Kolmogorov-Smirnov test statistic $d(1 - \alpha, n)$ to test for a normal distribution at different significance levels α

n	0.01	0.05	0.10	0.15	0.20
4	0.417	0.381	0.352	0.319	0.300
5	0.405	0.337	0.315	0.299	0.285
6	0.364	0.319	0.294	0.277	0.265
7	0.348	0.300	0.276	0.258	0.247
8	0.331	0.285	0.261	0.244	0.233
9	0.311	0.271	0.249	0.233	0.223
10	0.294	0.258	0.239	0.224	0.215
11	0.284	0.249	0.230	0.217	0.206
12	0.275	0.242	0.223	0.212	0.199
13	0.268	0.234	0.214	0.202	0.190
14	0.261	0.227	0.207	0.194	0.183
15	0.257	0.220	0.201	0.187	0.177
16	0.250	0.213	0.195	0.182	0.173
17	0.245	0.206	0.189	0.177	0.169
18	0.239	0.200	0.184	0.173	0.166
19	0.235	0.195	0.179	0.169	0.163
20	0.231	0.190	0.174	0.166	0.160
25	0.200	0.173	0.158	0.147	0.142
30	0.187	0.161	0.144	0.136	0.131
$n > 30$	$\dfrac{1.628}{\sqrt{n}}$	$\dfrac{1.358}{\sqrt{n}}$	$\dfrac{1.224}{\sqrt{n}}$	$\dfrac{1.138}{\sqrt{n}}$	$\dfrac{1.073}{\sqrt{n}}$

Digital filters

Table VII.

Coefficients for computing *first derivatives* (Savitzky and Golay, 1964)

Points	25	23	21	19	17	15	13	11	9	7	5
−12	−12										
−11	−11	−11									
−10	−10	−10	−10								
−9	−9	−9	−9	−9							
−8	−8	−8	−8	−8	−8						
−7	−7	−7	−7	−7	−7	−7					
−6	−6	−6	−6	−6	−6	−6	−6				
−5	−5	−5	−5	−5	−5	−5	−5	−5			
−4	−4	−4	−4	−4	−4	−4	−4	−4	−4		
−3	−3	−3	−3	−3	−3	−3	−3	−3	−3	−3	
−2	−2	−2	−2	−2	−2	−2	−2	−2	−2	−2	−2
−1	−1	−1	−1	−1	−1	−1	−1	−1	−1	−1	−1
0	0	0	0	0	0	0	0	0	0	0	0
+1	1	1	1	1	1	1	1	1	1	1	1
+2	2	2	2	2	2	2	2	2	2	2	2
+3	3	3	3	3	3	3	3	3	3	3	
+4	4	4	4	4	4	4	4	4	4		
+5	5	5	5	5	5	5	5	5			
+6	6	6	6	6	6	6	6				
+7	7	7	7	7	7	7					
+8	8	8	8	8	8						
+9	9	9	9	9							
+10	10	10	10								
+11	11	11									
+12	12										
NORM	1300	1012	770	570	408	280	182	110	60	28	10

Table VIII.

Coefficients for computing *second derivatives* (Savitzky and Golay, 1964)

Points	25	23	21	19	17	15	13	11	9	7	5
−12	92										
−11	69	77									
−10	48	56	190								
−9	29	37	133	51							
−8	12	20	82	34	40						
−7	−3	5	37	19	25	91					
−6	−16	−8	−2	6	12	52	22				
−5	−27	−19	−35	−5	1	19	11	15			
−4	−36	−28	−62	−14	−8	−8	2	6	28		
−3	−43	−35	−83	−21	−15	−29	−5	−1	7	5	
−2	−48	−40	−98	−26	−20	−48	−10	−6	−8	0	2
−1	−51	−43	−107	−29	−23	−53	−13	−9	−17	−3	−1
0	−52	−44	−110	−30	−24	−56	−14	−10	−20	−4	−2
+1	−51	−43	−107	−29	−23	−53	−13	−9	−17	−3	−1
+2	−48	−40	−98	−26	−20	−48	−10	−6	−8	0	2
+3	−43	−35	−83	−21	−15	−29	−5	1	7	5	
+4	−36	−28	−62	−14	−8	−8	2	6	28		
+5	−27	−19	−35	−5	1	19	11	15			
+6	−16	−8	−2	6	12	52	22				
+7	−3	5	37	19	25	91					
+8	12	20	82	34	40						
+9	29	37	133	51							
+10	48	56	190								
+11	69	77									
+12	92										
NORM	26910	17710	33649	6783	3876	6188	1001	429	462	42	7

Experimental designs

Table IX.

Two-level designs *(half-cell designs)* for 3, 4 and 5 factors

2^{3-1} design

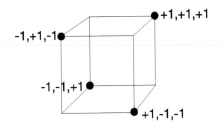

Run	Factors		
	x_1	x_2	x_3
1	−1	−1	+1
2	+1	−1	−1
3	−1	+1	−1
4	+1	+1	+1

2^{3-1} design

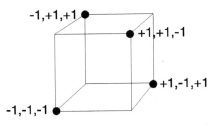

Run	Factors		
	x_1	x_2	x_3
1	−1	−1	−1
2	+1	−1	+1
3	−1	+1	+1
4	+1	+1	−1

2^{4-1} design

Run	Factors				Run	Factors			
	x_1	x_2	x_3	x_4		x_1	x_2	x_3	x_4
1	−1	−1	−1	−1	1	−1	−1	−1	+1
2	+1	−1	−1	+1	2	+1	−1	−1	−1
3	−1	+1	−1	+1	3	−1	+1	−1	−1
4	+1	+1	−1	−1	4	+1	+1	−1	+1
5	−1	−1	+1	+1	5	−1	−1	+1	−1
6	+1	−1	+1	−1	6	+1	−1	+1	+1
7	−1	+1	+1	−1	7	−1	+1	+1	+1
8	+1	+1	+1	+1	8	+1	+1	+1	−1

2^{5-1} design

Run	Factors				
	x_1	x_2	x_3	x_4	x_5
1	−1	−1	−1	−1	+1
2	+1	−1	−1	−1	−1
3	−1	+1	−1	−1	−1
4	+1	+1	−1	−1	+1
5	−1	−1	+1	−1	−1
6	+1	−1	+1	−1	+1
7	−1	+1	+1	−1	+1
8	+1	+1	+1	−1	−1
9	−1	−1	−1	+1	−1
10	+1	−1	−1	+1	+1
11	−1	+1	−1	+1	+1
12	+1	+1	−1	+1	−1
13	−1	−1	+1	+1	+1
14	+1	−1	+1	+1	−1
15	−1	+1	+1	+1	−1
16	+1	+1	+1	+1	+1

2^{5-1} design

Run	Factors				
	x_1	x_2	x_3	x_4	x_5
1	−1	−1	−1	−1	−1
2	+1	−1	−1	−1	+1
3	−1	+1	−1	−1	+1
4	+1	+1	−1	−1	−1
5	−1	−1	+1	−1	+1
6	+1	−1	+1	−1	−1
7	−1	+1	+1	−1	−1
8	+1	+1	+1	−1	+1
9	−1	−1	−1	+1	+1
10	+1	−1	−1	+1	−1
11	−1	+1	−1	+1	−1
12	+1	+1	−1	+1	+1
13	−1	−1	+1	+1	−1
14	+1	−1	+1	+1	+1
15	−1	+1	+1	+1	+1
16	+1	+1	+1	+1	−1

Table X.

Central composite design for four factors with triplicate measurements in the center of the design

Run	Factors			
	x_1	x_2	x_3	x_4
1	1	−1	1	−1
2	−1	−1	1	1
3	−α	0	0	0
4	1	1	−1	1
5	0	0	0	0
6	−1	1	1	1
7	0	0	0	α
8	−1	1	1	−1
9	0	−α	0	0
10	−1	−1	1	−1
11	0	0	0	0
12	−1	−1	−1	1
13	1	−1	1	1
14	0	0	α	0
15	−1	1	−1	−1
16	1	−1	−1	1
17	α	0	0	0
18	1	1	1	1
19	−1	1	−1	1
20	1	1	1	−1
21	1	1	−1	−1
22	0	0	0	0
23	0	α	0	0
24	−1	−1	−1	−1
25	1	−1	−1	−1
26	0	0	0	−α
27	0	0	−β	0

Table XI.

Box-Behnken design for four factors with triplicate
measurements in the center of the design

Run	Factors			
	x_1	x_2	x_3	x_4
1	0	0	−1	1
2	1	1	0	0
3	0	−1	0	1
4	0	0	−1	−1
5	−1	1	0	0
6	0	−1	0	−1
7	1	0	1	0
8	−1	−1	0	0
9	1	0	0	1
10	0	−1	1	0
11	0	1	−1	0
12	0	0	1	−1
13	1	0	−1	0
14	1	0	0	−1
15	1	−1	0	0
16	0	0	1	1
17	−1	0	0	1
18	0	0	0	0
19	0	1	1	0
20	0	1	0	−1
21	−1	0	−1	0
22	0	1	0	1
23	−1	0	0	−1
24	0	0	0	0
25	0	0	0	0
26	0	−1	−1	0
27	−1	0	1	0

Table XII.

Mixture designs *(lattice designs)* for three and four factors

Run	Factor		
	x_1	x_2	x_3
1	1	0	0
2	0.67	0.33	0
3	0.67	0	0.33
4	0.33	0.67	0
5	0.33	0.33	0.33
6	0.33	0	0.67
7	0	1	0
8	0	0.67	0.33
9	0	0.33	0.67
10	0	0	1

Run	Factor			
	x_1	x_2	x_3	x_4
1	1	0	0	0
2	0.67	0.33	0	0
3	0.67	0	0.33	0
4	0.67	0	0	0.33
5	0.33	0.67	0	0
6	0.33	0.33	0.33	0
7	0.33	0.33	0	0.33
8	0.33	0	0.67	0
9	0.33	0	0.33	0.33
10	0.33	0	0	0.67
11	0	1	0	0
12	0	0.67	0.33	0
13	0	0.67	0	0.33
14	0	0.33	0.67	0
15	0	0.33	0.33	0.33
16	0	0.33	0	0.67
17	0	0	1	0
18	0	0	0.67	0.33
19	0	0	0.33	0.67
20	0	0	0	1

Matrix algebra

A point in n-dimensional space R^n is represented by a *vector*, i.e.

$$x = \begin{pmatrix} x_1 \\ x_2 \\ \vdots \\ x_n \end{pmatrix} \quad \text{or in transposed form} \quad x^T = (x_1, x_2, \ldots, x_n)$$

The sum of two vectors $x, y \in R^n$ gives:

$$x + y = \begin{pmatrix} x_1 + y_1 \\ x_2 + y_2 \\ \vdots \\ x_n + y_n \end{pmatrix} \quad \text{example:} \quad \begin{pmatrix} 1 \\ 2 \\ 3 \end{pmatrix} + \begin{pmatrix} 2 \\ 4 \\ 7 \end{pmatrix} = \begin{pmatrix} 3 \\ 6 \\ 10 \end{pmatrix}$$

Multiplication of a vector x with a scalar $l \in R^n$ gives the following vector:

$$l \times x = \begin{pmatrix} lx_1 \\ lx_2 \\ \vdots \\ lx_n \end{pmatrix}$$

A *matrix* of elements of real numbers consisting of n rows and m columns, i.e. a $n \times m$ – matrix, is defined by:

$$A = \begin{pmatrix} a_{11} & a_{12} & \cdots & a_{1m} \\ a_{21} & a_{22} & & a_{2m} \\ \vdots & & & \vdots \\ a_{n1} & a_{n2} & \cdots & a_{nm} \end{pmatrix} \quad \text{example:} \quad A = \begin{pmatrix} 2 & 4 & 6 \\ 3 & 1 & 5 \\ 5 & 8 & 9 \end{pmatrix}$$

A *square matrix* has the same number of rows and columns, i.e. its dimensions are $n \times n$.

In a transposed matrix A^T, the rows and columns are interchanged giving for the matrix A:

$$A^T = \begin{pmatrix} a_{11} & a_{21} & \cdots & a_{n1} \\ a_{12} & a_{22} & & a_{n2} \\ \vdots & & & \vdots \\ a_{1m} & a_{2m} & \cdots & a_{nm} \end{pmatrix} \quad \text{example:} \quad A^T = \begin{pmatrix} 2 & 3 & 5 \\ 4 & 1 & 8 \\ 6 & 5 & 9 \end{pmatrix}$$

If the transpose of a matrix is identical with the original matrix in every element, i.e. $A^T = A$, it is called a *symmetric matrix*.

A *diagonal matrix* is a special case of a symmetric matrix. In a diagonal matrix only the main diagonal contains values different from zero and all off-diagonal elements are zero:

$$A = \begin{pmatrix} a_{11} & 0 & \cdots & 0 \\ 0 & a_{22} & & 0 \\ \vdots & & & \vdots \\ 0 & 0 & \cdots & a_{nn} \end{pmatrix} \qquad \text{example: } A = \begin{pmatrix} 2 & 0 & 0 \\ 0 & 1 & 0 \\ 0 & 0 & 9 \end{pmatrix}$$

The diagonal matrix that has all 1s on the diagonal is termed the *identity matrix*:

$$I = \begin{pmatrix} 1 & 0 & \cdots & 0 \\ 0 & 1 & & 0 \\ \vdots & & & \vdots \\ 0 & 0 & \cdots & 1 \end{pmatrix}$$

The following examples illustrate *matrix addition* and *matrix subtraction*:

$$A + B = \begin{pmatrix} 2 & 4 \\ 1 & 3 \end{pmatrix} + \begin{pmatrix} -1 & 2 \\ 5 & -3 \end{pmatrix} = \begin{pmatrix} 1 & 6 \\ 6 & 0 \end{pmatrix}$$

$$A - B = \begin{pmatrix} 2 & 4 \\ 1 & 3 \end{pmatrix} - \begin{pmatrix} -1 & 2 \\ 5 & -3 \end{pmatrix} = \begin{pmatrix} 3 & 2 \\ -4 & 6 \end{pmatrix}$$

Multiplication of an $n \times n$ matrix A and an $n \times n$ matrix B gives the $n \times n$ matrix C:

$$C = AB = \begin{pmatrix} a_{11} & \cdots & a_{1k} \\ \vdots & & \\ a_{n1} & \cdots & a_{nk} \end{pmatrix} \begin{pmatrix} b_{11} & \cdots & b_{1m} \\ \vdots & & \\ b_{k1} & & b_{km} \end{pmatrix} = \begin{pmatrix} c_{11} & \cdots & c_{1m} \\ \vdots & & \\ c_{n1} & & c_{nm} \end{pmatrix}$$

where $c_{ij} = \sum_{l=1}^{k} a_{il} b_{lj}$ for $1 \le i \le n$ and $1 \le j \le m$.

Example: $C = \begin{pmatrix} 2 & 3 & 1 \\ 3 & 4 & 1 \end{pmatrix} \begin{pmatrix} 7 & 4 \\ 1 & 3 \\ 5 & 0 \end{pmatrix} = \begin{pmatrix} 22 & 17 \\ 30 & 24 \end{pmatrix}$

The *rank* of a matrix is the maximum number of linearly independent vectors (rows or columns) in an $n \times p$ matrix X denoted as $r(X)$. Linearly dependent rows or columns reduce the rank of a matrix.

The *determinant* of a matrix is calculated from:

$$D = \begin{vmatrix} a_{11} & a_{12} & \cdots & a_{1m} \\ a_{21} & a_{22} & & a_{2m} \\ \vdots & & & \\ a_{n1} & a_{n2} & \cdots & a_{nm} \end{vmatrix} = \sum_{i=1}^{n} (-1)^{\text{ith}} \, a_{ik} \det(M_{ik})$$

Here M_{ik} is the $(n-1) \times (n-1)$ matrix where the ith row and kth column has been deleted.

Example: $D = \begin{vmatrix} 2 & 4 & 6 \\ 3 & 1 & 5 \\ 7 & 8 & 9 \end{vmatrix} = 2(1 \cdot 9 - 5 \cdot 8) - 4(3 \cdot 9 - 5 \cdot 7) + 6(3 \cdot 8 - 1 \cdot 7) = 72$

For *inversion* of a matrix A we get A^{-1}, where:

$$A = \begin{pmatrix} a_{11} & a_{12} \\ a_{21} & a_{22} \end{pmatrix} \qquad A^{-1} = \begin{pmatrix} \dfrac{a_{22}}{D} & -\dfrac{a_{12}}{D} \\ -\dfrac{a_{21}}{D} & \dfrac{a_{11}}{D} \end{pmatrix}$$

where D is the determinant of the matrix.

Example: $A = \begin{pmatrix} 4 & 2 \\ 3 & 1 \end{pmatrix} \quad A^{-1} = \begin{pmatrix} -0.5 & 1 \\ 1.5 & -2 \end{pmatrix}$

In this example we have inverted a 2×2 matrix. Perhaps mental inversion could also be performed in the case of a 3×3 matrix. For larger matrices, however, a computer algorithm is necessary. In addition, matrix inversion is a very sensitive procedure, so that powerful algorithms, such as singular value decomposition (cf. Sec. 5.2.1) must be applied.

A *linear transformation* from R^n to R^m (case a) or from R^m to R^n (case b) is possible by:

a. multiplication of a n-dimensional column vector by an $n \times m$-matrix forming a m-dimensional vector:

$$x^T A = (x_1, x_2, \ldots x_n) \begin{pmatrix} a_{11} & \cdots & a_{1m} \\ a_{21} & & a_{2m} \\ \vdots & & \\ a_{n1} & \cdots & a_{nm} \end{pmatrix} = \left(\sum_{i=1}^{n} x_i a_{i1}, \ldots, \sum_{i=1}^{n} x_i a_{im} \right)$$

b. multiplication of an $n \times m$-matrix by a m-dimensional row vector:

$$Ax = \begin{pmatrix} a_{11} & \cdots & a_{1m} \\ a_{21} & & a_{2m} \\ \vdots & & \\ a_{n1} & \cdots & a_{nm} \end{pmatrix} \begin{pmatrix} x_1 \\ x_2 \\ \vdots \\ x_m \end{pmatrix} = \begin{pmatrix} \sum_{i=1}^{m} x_i a_{1i} \\ \sum_{i=1}^{m} x_i a_{2i} \\ \vdots \\ \sum_{i=1}^{m} x_i a_{ni} \end{pmatrix}$$

305

Vectors that do not change their direction during a linear transformation are very important. They are termed the *eigenvectors* of the matrix A. For every eigenvector of A there exists a real number λ, the eigenvalue, for which the following equation is valid:

$$Ax = \lambda x$$

Eigenvector analysis is needed, e.g. in Sec. 5.2.1 for projection of multidimensional data.

Software

General statistics

Statgraphics
(for Windows), Manugistics, Inc.

Statistica
(for Windows), StatSoft Inc.

Excel (Analysis functions)
(for Windows), Microsoft Inc.

MINITAB Statistical Software
MINITAB Inc.

SPSS
(for Windows), SPSS Inc.

Chemometrics

SCAN – **S**oftware for **C**hemometric **An**alysis
Minitab Inc.

Unscrambler
CAMO – Computer Aided Modelling A/S

MathLab with chemometric toolbox
The MatWorks Inc.

GRAMS/32
(for Windows), Galactic Industries

Neural Networks

Neuralworks
NeuralWare Inc.

Index

Factor
   coding of 108
   common 133
   experimantal 81
   significance of 142
   specific 134
Factor analysis 133
   evolving 148
Factor effect
   interaction 97
   main 97
   relation with regression parameters 109
Factorial
   experiment 92
   methods 124 ff.
Feature 121
Figure-of-merit *see* objective function
Filter
   high pass 66
   Kalman 56 ff.
   low pass 66
   moving average 52
   polynomial 53 ff.
   recursive 56
   Savitzky-Golay 53
   width 52, 54 ff.
Forecasting *see* prediction
Fragmentation code 231
Frame 247
Frequency 15 ff.
   aliased 63
   function 15 ff.
Fuzzy set 268 ff.
   operation 273
Fuzzy theory 11, 267

Gaussian distribution *see* normal distribution
Generalized inverse 193
Generic structure *see* Markush structure
Genetic algorithm 276 ff.
Good laboratory practice 288
Goodness-of-fit test 183
Graph theory 232
Grubb's test 40

Hamming distance 240
Hardware 4
Hat-matrix 208 ff.
Header 229
Hebb learning *see* learning, associative

Heridity 276
Hess's matrix 219
Hexadecimal number 4
Hidden layer 254
Histogram 14 ff., 189
Hit list 238
HORD code 232
Horn clause 250
HOSE code 231
Hypothesis
   alternative 28 ff.
   null 28 ff.
   testing 28

I/O system 6
Identity matrix 304
Indicator function
Induction 249
Inference 248
   engine 252
Influential observation 210
Information theory 10
Interlaboratory comparison 287
Interquartile range 21
Intersection 273
Inverted list 237

Jackknifed residual *see* Studentized residual
Jacobian matrix 218

Knowledge
   base 252
   processing 245
   representation 246
Kohonen network 265
Kurtosis *see* distribution

Laboratory-information-and-management
   system 7, 236
Lack 283
Lack-of-fit test 183
Latent variable 126, 197
Latin square *see* experimental design
Learning
   associative 259
   competitive 259
   correlation 256
   supervised 160 ff., 258, 260
   unsupervised 124 ff., 258, 260
Learning coefficient 259
Learning law 259